T0094364

VEGETAL SEX

ALSO AVAILABLE FROM BLOOMSBURY

Dump Philosophy: A Phenomenology of Devastation, Michael Marder

General Ecology: The New Ecological Paradigm, ed. Erich Hörl with James Burton

Posthuman Glossary, ed. Rosi Braidotti and Maria Hlavajova

VEGETAL SEX

Philosophy of Plants

Stella Sandford

BLOOMSBURY ACADEMIC
LONDON • NEW YORK • OXFORD • NEW DELHI • SYDNEY

BLOOMSBURY ACADEMIC
Bloomsbury Publishing Plc
50 Bedford Square, London, WC1B 3DP, UK
1385 Broadway, New York, NY 10018, USA
29 Earlsfort Terrace, Dublin 2, Ireland

BLOOMSBURY, BLOOMSBURY ACADEMIC and the Diana logo are
trademarks of Bloomsbury Publishing Plc

First published in Great Britain 2023

Cover design by Ben Anslow
Cover image: Thorn Bush Branch (© shanelinkcom / iStock)

A catalogue record for this book is available from the British Library.

A catalog record for this book is available from the Library of Congress.

ISBN: HB: 978-1-3502-7492-1
         PB: 978-1-3502-7493-8
         ePDF: 978-1-3502-7494-5
         eBook: 978-1-3502-7495-2

Typeset by Integra Software Services Pvt. Ltd.

To find out more about our authors and books visit www.bloomsbury.com
and sign up for our newsletters.

For my offsprings.

Much do they teach; long may they love and thrive.

CONTENTS

ACKNOWLEDGEMENTS

This book was mostly written under the restricted conditions made necessary by the coronavirus pandemic that began in early 2020.

I thank all the staff at the London Local Authority Public Library Network who safely reopened the library service when it became possible and sent books for me to pick up from my local libraries. I also thank the staff at the Linnean Society library and the British Library.

I am grateful for the support, encouragement and criticism from colleagues and students, past and present, at the Centre for Research in Modern European Philosophy at Kingston University, London; from some treasured intellectual co-workers and friends, especially Lisa Baraitser, Erica Fudge and Tuija Pulkkinen; and from my friend and walking partner, Josephine Grosset. I also thank Liza Thompson at Bloomsbury for her enthusiasm for this project and Daniel Whistler for his intellectual support and championing of plant philosophy. Special thanks are owed to Peter Osborne, for quite a lot of things, intellectual and otherwise.

I am also thankful for the space and vegetable life of Abney Park Cemetery, which helped me to think, and where I remember Maureen Ivy Deverall, née Lynn, 1937–2019.

The research for this book was made possible by a Major Research Fellowship from the Leverhulme Trust ('Sex Difference in Natural History', MRF 2017-006). I am very grateful to the Leverhulme Trust for this generous funding.

INTRODUCTION

What is vegetal sex? Or, what does it mean to say that plants are sexed? Do 'male' and 'female' straightforwardly mean the same when applied to humans, trees, fungi and algae? Given that the discovery of the sex of plants is taken to be a major step in the development of scientific botany, it might seem regressive to ask these questions. That plants are sexed is, in most respects, simply taken for granted today by most plant scientists, horticulturalists and amateur gardeners alike. It is also generally taken for granted that the progress of scientific botany depended on its freeing itself from proceeding through analogies between plants and animals, where the animal always provided the model, and on its freeing itself from the mystifications of philosophy. As these plant-animal analogies were most developed in philosophical texts on plants, these two constraints on scientific progress are, indeed, seen as inseparable. But how is it, then, that in one exceptionally important case – the discovery of the sex of plants – progress is measured not by the rejection of the animal model but, on the contrary, by its total and literal embrace? What is the significance of the fact that the discovery of the sex of plants thus represents a distinct anomaly in relation to the general methodological tendency in scientific botany? What is the significance of the fact that the animal model still dominates when it comes to plant sex?

This book addresses these questions through detailed analyses of emblematic moments in the history of the theory of plant sex. It does this from the standpoint of contemporary plant philosophy while also re-evaluating the contribution of the longer tradition of plant philosophy to the question of plant sex. For this tradition, the investigation of the question of whether plants were divided into male and female required a prior, philosophical question to be asked: What are 'male' and 'female'? The aim of this book is to show that the problem of the meaning of plant sex today (because that meaning cannot, in fact, be taken for granted) still requires us to ask this prior question, and thus gives us human beings a stake in the question of plant sex that goes beyond our understanding of plants. It also aims to show that the much-trumpeted liberation of botany from philosophy may be less complete – and is certainly less desirable – than the traditional histories of botany assume.

What, then, is the standpoint of contemporary plant philosophy? The plant philosophy that has arisen in recent decades, mainly in a Euro-American context, is related to, but distinct from, the recent crop of popular books on plants (especially trees) and fungi that advocate for a better understanding of the complexities of plant and fungal life, often by ascribing to them attributes or forms of behaviour previously reserved for animals (e.g. motion, sensitivity, perception) or reserved more particularly for humans (e.g. intelligence, consciousness). Together, plant philosophy and plant and fungal advocacy contest what they see as a general under-estimation of plants and a cultural failure to take plants seriously, in several respects. According to writers in these fields, we simply do not pay enough attention to plants. We take them for granted, objectify and 'exploit' them, notably in their reduction to agro-capitalist commodities. As plant life stands metonymically for 'nature' as a whole, the devaluation and commodification of plants is at the same time the devaluation and commodification of nature as a whole, which has contributed to the capitalist-driven global environmental catastrophe that faces us today. Plant philosophy, further, finds an explanation and justification for this devaluation of plant life in the foundational categories of Western philosophy, or finds that devaluation to be part of the foundation of that metaphysics.

What we are calling 'plant advocacy' is a popular scientific genre with contributions from plant scientists and those who work professionally with plants (such as foresters) and from plant enthusiasts. Plant philosophers read this plant advocacy literature, but plant advocates are generally less keen on plant philosophy, judging by their bibliographies. Plant advocates contest what they see (no doubt correctly) as the prevailing relegation of plants to second class in relation to animal life. They thus contest one kind of 'zoocentrism', that is, the tendency to assume the superiority of animal life over all other forms. Against this kind of moral zoocentrism (as we shall call it) plant advocates stress, amongst other things, the extent of plant dominance on earth (estimates vary but at least 80 per cent of all biomass on the planet is thought to be vegetal; recent estimates go as high as 99 per cent) and the total reliance of all animal life on plants and fungi. (This is not to overlook the role of bacteria, but that is another story.) However, the rhetoric of plant advocacy is itself often thoroughly zoocentric and anthropomorphic in another sense when it urges us to recognize that plants, too, have (at least some) of the capacities and attributes that animals and humans have.

Plant philosophy, on the other hand, is often explicitly phytocentric, or at least aims to be. Some plant philosophy aspires to an 'ontophytology'[1] – an ontology of plant being – and argues for the recognition of the vegetative aspects of animal being. It also sometimes aspires to a vegetal ethics – both an attempt to include plant life within the sphere of ethical consideration (so that agro-capitalism becomes an ethical matter for its treatment of plants and not just because of the fallout for humans, although the two are related) and an attempt to articulate an ethics that begins from the vegetal (once 'the vegetal' has been understood from a phytocentric perspective). While this is not the attempt to philosophize *as* plant, at its further reaches it is an attempt to philosophize from vegetal being.

As Quentin Hiernaux and Benoît Timmermans say, the field of plant philosophy does not present a unified front, let alone comprise a system.[2] But some common themes do emerge. In his overview of the field Hiernaux identifies three common tendencies in attempts to philosophize on the vegetal, which are all forms of intellectual resistance to traditional conceptions of the living being (*le vivant*): resistance to animal models, resistance to the idea of the organism as an individual and resistance to the duality organism/environment.[3] It is the being of plants themselves that resist these traditional conceptions of the living being, in the sense that the latter are inadequate to the specificity of vegetal being. This leads Michael Marder to the strong claim that 'in its very being the plant accomplishes a lived destruction of metaphysics… carrying out a transvaluation of metaphysical values'.[4] For Marder the ontology of plant being must therefore be 'post-metaphysical'.[5]

This book is intended as a contribution to plant philosophy, understood in the broadest sense. It builds on the achievement and insights of existing plant philosophy and plant thinking but also presupposes that plant philosophy requires critical investigation of aspects of the history of botany and of the modern plant sciences, as opposed to attempts to determine the philosophical contours of vegetal being from a purely analytic perspective. It thus approaches the history of botany and the modern plant sciences with a philosophical eye, and in that sense looks at botany through philosophy. But throughout, the book also aims to demonstrate some of the ways in which philosophy and botany have been and still are inextricably entwined, for example in botany's tacit reliance on philosophical categories or in some modern botanical works that develop, despite themselves perhaps, in philosophical ways.

In this book the primary 'resistance', to use Hiernaux's word, is to the assumption of animal models in attempts to describe and characterize aspects of vegetal being. More particularly, it is concerned with the explicit and implicit use of analogies between plants and animals (the tendency to understand plants via an analogy with animals) in one specific area: vegetal sex. It is surprising that this is a topic that has hitherto largely been absent from plant philosophy, because there is probably no aspect of plant science in which the animal model was and remains more influential and in which the animal model continues to function more uncritically than in the ways that we speak about plant sex and sexual reproduction, even today.[6] But in extending contemporary plant philosophy to include the specific problem of the meaning of vegetal sex – what 'male' and 'female' have meant and can mean in relation to plant life – the book also raises more general questions about the generic concepts of 'sex', 'male' and 'female'.

Nothing in this book aims to contest or question the findings of the modern plant sciences; we take these sciences on trust. Rather, it is our intention to question the language and conceptual presuppositions that shape some aspects of these sciences and the communication of their findings in the scientific and public-facing literature. To this end we ask: How has plant sex been understood in the history of botany? According to what models? How has the analogy with animal sex shaped and delimited what it is possible to say about plant sex and sexual reproduction? How does it continue to shape that discourse? How do plant scientists characterize plant sex today, and how does this differ from the characterizations of earlier scientific eras? But also, how does the characterization of sex in modern plant science still echo the presuppositions of earlier eras? How does some of that science strive, still, to free itself from the animal model and the conceptual forms to which that model is applied? Finally, if the animal model that has dominated discussions of vegetal sex is inadequate to it – as we believe that we can show that it is – what does this mean for the generic concept of sex? Is there a meaningful generic concept of sex (or of male and female), and if so to what, exactly, does it apply?

Towards the end of his *Éloge de la plante*, the botanist Francis Hallé writes:

> I dream of a botany that could determine itself autonomously, according to its own rules, no longer lagging behind animal or human physiology. Thinking about the plant itself, as an original form of life, as a model with respect to autonomy and the restoration of the environment, botany could regain its place at the centre of the life sciences.[7]

The questions tackled in this book can be summed up in relation to Hallé's dream: Has botany ever determined itself autonomously, according to its own rules, when it comes to plant sex? The investigation of this question is a task for a plant philosophy that takes the plant sciences for its object, not a task for plant science itself.[8] As plant philosophy is not science, its categories are philosophical, not scientific. Whereas, scientifically, it would usually be thought right to distinguish between plants, fungi and algae,[9] this book will work with the philosophical category of the 'plant' or the 'vegetal', encompassing all three of those groups. Vegetal philosophy is concerned with conceptual, not evolutionary, phylogeny.

Chapter 1 sets out the contemporary philosophical context for the book in more detail, introducing the field of plant philosophy via an argument about its distinction from plant advocacy. Explaining the specificity of plant philosophy, establishing the field from which our investigations set out, will also allow us to see more easily what has hitherto been missing from it: an investigation of the specificity of vegetal sex. The chapter begins with a critical discussion of some of the claims in popular plant advocacy about the senses of plants (plant 'seeing' and 'hearing', in particular), with reference to the science on which these claims are based. It argues that while the advocates of plant sensing pit themselves against a kind of *moral* zoocentrism (they are concerned to show that plants are not inferior to animals in complexity and that they deserve our respect), plant advocacy is itself zoocentric in another sense, because it remains reliant on animal models. But what if we were to resist the tendency to think of complex plant behaviours in zoocentric ways as 'seeing', 'hearing' and so on? What possibilities are there for recognizing the plant-specific 'senses' that go beyond the animal varieties?

The plant advocacy literature is also often concerned with the idea of plant intelligence primarily associated with the work of the plant behaviourist Anthony Trewavas. As in its discussions of plant sensing, the plant advocacy literature tends to interpret this idea anthropomorphically. However, this chapter argues for an interpretation of Trewavas's work on plant intelligence in terms of his specific (albeit implicit) philosophical approach. Although Trewavas and those influenced by him do not avoid zoocentric models altogether, this chapter argues that a philosophical interpretation of the idea of plant intelligence shows it to be essentially anti-zoocentric, displacing the equation of intelligence in general with its specifically human forms. This argument thus allies Trewavas's conception of plant intelligence more with plant philosophy than with plant advocacy, his influence on the latter notwithstanding.

The second part of Chapter 1 then concentrates on the more explicitly anti-zoocentric approach of plant philosophy. Plant philosophy, we argue, attempts to think plant life outside of zoocentric models, and this is particularly clear in the work of the botanist-philosophers Francis Hallé and Jacques Tassin. The features that plant philosophers identify as specific to plant life (their modular principle of organization, lack of central control, phenotypic plasticity and so on) are already axioms of the plant sciences. But plant philosophy attempts to draw out the philosophical significance of these features through a confrontation of them with the kinds of metaphysical categories (e.g. that of the individual) that we tend to take for granted in thinking about life. What does vegetal resistance to these categories suggest to us about their philosophical status? The more challenging plant philosophy of thinkers such as Michael Marder and Emanuele Coccia pushes us even further. What if the plant were to become, in some sense, a model for aspects of animal – and specifically human – life? Do we animals have a vegetal part?

Acknowledging the insights of existing plant philosophy, Chapter 1 nevertheless ends by posing a major question to it: What about plant sex? When plant philosophers and botanists alike stress the specificity of plant being by insisting that they do not have differentiated organs like animals, why do they nevertheless tend to speak freely of the sex organs of plants? Given that plants alone, of all living forms, have a dibiontic life cycle, why has plant philosophy and thinking had so little to say about this? Why does the carrying over of terminology from zoology ('male', 'female', 'mother', 'sister', 'parent', 'offspring') into contemporary talk of plants not raise any eyebrows in plant philosophy or botany? Can we be sure that the thoroughly anthropomorphic terminology used to describe plant reproduction and relations has no effect on scientific understanding?

Can we assume that 'sex' means the same in relation to both animals and plants? And if 'sex' and more particularly 'male' and 'female' do not mean the same thing in relation to both plants and animals, what becomes of our generic concept of 'sex'? What is implied concerning the need to specify what we mean (and we may mean many things) when we refer to ourselves as 'male' and 'female'?

There is an historical irony in the fact that the new plant philosophy has shied away from questioning the adequacy of the animal model of sex for plants, because this was a major part both of earlier traditions of plant philosophy and of the history of the scientific theory of plant sexuality: that is, the history of the ascription of sex (male and female) to plants. Indeed, these two fields were, until well into the eighteenth century, inseparably entwined, to the extent that, for most of this history, the science of plant sexuality *is* plant philosophy. We thus begin our investigation of vegetal sex with a series of studies of emblematic moment in the philosophical history of the theory of plant sexuality.

Because the discovery of plant sexuality is taken to be a major development in scientific botany, all general histories of botany include a discussion of it. In these traditional histories of botany (as we shall call them) three presuppositions frame the interpretation of this discovery. First, writing from the standpoint of modern scientific knowledge (at whatever stage that had achieved for the respective writers) the traditional histories of botany tend to take the idea of plant sexuality for granted (as if plant sex is now obvious) and to search for early indications of it from ancient Greece to the eighteenth century. Often, it proceeds in an explicitly Whiggish fashion and castigates earlier thinkers for failing to discover plant sexuality before the eighteenth century. This is true, for example, of the most detailed existing history of the theory of plant sexuality, Lincoln Taiz and Lee Taiz's *Flora Unveiled*, which opens with 'two basic questions: why did it take so long to discover sex in plants, and why, after its proposal and experimental confirmation in the late seventeenth century, did the debate continue for another 150 years?'[10] Second, it is presumed that the inauguration and progress of scientific botany are dependent on the liberation of botany from philosophy, which latter could only act as a form of mystification or an obstacle to scientific thought. Early plant philosophy, following Aristotle, often worked through a series of analogies with animals, a situation from which scientific botany struggled to free itself. The third presupposition in the traditional histories of botany is that the liberation of botany from philosophy, around the end of the seventeenth and the beginning of the eighteenth centuries, also, in the case of the theory of plant sexuality at least, releases botany from the mere analogy with animal sex, allowing for an understanding of plants sex as literally – not just metaphorically – sexed.

This book contests all three of these presuppositions; Chapter 2 deals with the first and second of them particularly. The written history of the theory of plant sexuality in the West effectively begins with Aristotle, whose influence determined the course of plant philosophy until well into the sixteenth century and remained exceptionally strong for centuries after. Chapter 2 begins the investigation of this history via an analysis of the function of analogy in Aristotle's philosophical zoology and the specific analogy with animals that informed answers to the question 'Do plants have "male" and "female"?' We show that this question could not be addressed without first asking another, more fundamental philosophical question – 'What are "male" and "female"?' We show that there are at least three distinct meanings of 'male' and 'female' in Aristotle and the Aristotelian botanical tradition: 'male' and 'female' as metaphysical principles, 'male' and 'female' as individuals and 'male' and 'female' as parts of individuals or organs. Reconnecting twenty-first-century plant philosophy with a much longer tradition, this chapter thus shows how it is in thinking about plants that we come to see the difficulty of the basic question of the commonality of the meaning of 'male' and 'female' across different domains of life. We suggest, finally, that Aristotle's distinctions between the different meanings of 'male' and 'female', which we see structuring the plant philosophy of some of his most important successors, might well shed light on some of the confusions and ambiguities that tend to bedevil discussions of sex (in all kingdoms of life) today.

As we have said, traditional histories of botany tend to take the meaning of sex and sexuality for granted, and to proceed from the presupposition that plant sexuality is now an obvious fact. But according to these narratives this could only become obvious when botany freed itself from its Aristotelian baggage, or from philosophy itself. With the invention of microscopy and the shift (as the traditional histories see it) from speculation to experiment, the emergence of plant anatomy and physiology in the seventeenth century is seen as the beginning of a properly scientific botany. One of the greatest achievements of this era of botany was the 'discovery' of plant sex. Chapter 3 concentrates on this 'breakthrough' moment in the history of the theory of plant sexuality, with a detailed examination of the relevant work of the botanist to whom the discovery of the sex of plants is often attributed: Nehemiah Grew.

One sentence from Grew's *The Anatomy of Plants* (1682), often understood to be an identification of the plant stamen with the animal penis, is usually taken to mark the decisive beginnings of the scientific (as opposed to the metaphorical or speculatively philosophical) recognition of plant sex. But putting Grew's famous sentence back in its context in *The Anatomy of Plants* and re-uniting Grew's philosophy (especially his *Cosmologia Sacra*, 1701, and various philosophical essays) with his botany, Chapter 3 contests the idea that his comments on the sex of plants are the result of a rejection of philosophical deduction in favour of empirical observation. This chapter argues instead that Grew's comments on plant sex should be understood in the context of his philosophy of life and his attempt to understand plant generation according to a broadly Aristotelian animal model. It shows that philosophical speculation and a metaphysical conception of 'male' and 'female' still play a formative role in Grew's scientific 'discovery' of sex in plants (as it also does in John Ray's *Historia Plantarum Generalis*, 1686), and thus that plant philosophy – the joint venture of philosophy and botany – continued to be productive in the scientific era.

Chapter 4 concludes the historical part of our investigation by addressing two paradoxes in the traditional history of the theory of plant sexuality. First, in the literature on the use of analogy (or analogical reasoning) in the sciences it is generally agreed that, while analogy is a useful method to guide discovery, it always carries with it the danger that it will be pushed too far. More particularly, analogy is liable to be abused when the similarities that it tracks across different domains are misinterpreted as identities. Uniquely, however, the greatest scientific progress in the history of the theory of the sexuality of plants is, on the contrary, said to be marked precisely by the *identification* of the terms of the analogy between animals and plants, when the allegedly literal (not metaphorical) truth of plant sex is recognized. Second, the most obvious feature of the texts credited with popularizing this scientific step – the recognition of the literal truth of plant sex – is their *literary* (not literal) form, being structured according to extended analogies between the parts and behaviours of (human) animals and plants and a thoroughgoing metaphorization of plant sex. How can that be explained?

In this chapter we examine this move from analogy to identity in Rudolf Jakob Camerarius's *Letter on the Sex of Plants* (1694) and its relation to Aristotelian plant philosophy and raise the problem (addressed at greater length in Chapter 5) of precisely what in plant life is identified as 'literally' male and female. We then look at the eighteenth-century popularization of the idea of plant sex in Sébastien Vaillant and Carl Linnaeus. Traditional histories of botany try to detach the expansive sexual metaphorics of Vaillant's and Linnaeus's accounts of the nuptials of plants from the literal scientific truth of plant sex. This chapter argues, to the contrary, that these two aspects are inseparable in Vaillant's and Linnaeus' texts, as we see particularly in the construction of Linnaeus' scientific, botanical terminology through an extended analogy with animals and with aspects of human social arrangements. This, we argue, is made possible by a compounding or confusing of the different senses of 'male' and 'female' that were distinguished in the Aristotelian plant philosophy tradition, leading Vaillant and Linnaeus to a popular personification of plant sexes that has left its mark in the language of the plant sciences today. This suggests, we conclude, that we still need to ask what we mean by 'male' and 'female' in plants – a question that is as much philosophical as it is botanical.

In Chapter 5 we see how this question was indeed raised again after the discovery of the dibiontic life cycle unique to plants. This discovery gave rise, at the end of the nineteenth century and the beginning of the twentieth, to a terminological debate concerning the use of the terms 'male' and 'female' in relation to plants. In this chapter, then, we outline the nature of the dibiontic life cycle of plants and the subsequent debates about plant sex, arguing that this allows the full complexity of the problem of plant sex to emerge. We show how the idea of a popular 'untechnical' vocabulary for plant sex was admitted into the plant sciences and the extent of its use to the present day. We then look at what we argue is the most sophisticated attempt in the plant sciences to date to develop a specifically vegetal concept of plant sex, in the work of the botanist David G. Lloyd. We argue that Charles Darwin's puzzled comments on the sexes of plants, in his *The Different Forms of Flowers* (1877), reveal the need for a plant model for plant sex, and that this is what Lloyd attempts to provide. Lloyd's work is widely cited in the scientific literature and his standing as a botanist is well established. But this chapter argues that Lloyd's work on plant 'gender' can also be interpreted as proposing a philosophical account of plant sex, in the form of a relational epistemology in which the terms 'male', 'female', 'sex' and 'gender' are reconceptualized for plants, with definitions that are at odds with their animal meanings.

The final chapter (Chapter 6) concentrates on a recent development in thinking about forests – Suzanne Simard's idea of the 'Mother Tree'. Connecting this up with the earlier identification of the plant embryo and the plant placenta, we argue that these bear witness to what we call a 'maternal botanical imaginary' (in Michèle Le Doeuff's sense of 'imaginary', as we explain), in which the social and ideological aspects of the popular discussion of plant sex are arguably most clear. But Simard's idea of the Mother Tree also brings to the fore the difficult but essential question of the relationship between local or indigenous knowledge[11] on the one hand and Western science and philosophy on the other, as we see discussed in works by Robin Wall Kimmerer and Dale Turner. Simard's idea of the Mother Tree is obviously anthropocentric in some sense. Thinking about the relations between the scientific idea of the Mother Tree and its place in much older systems of thought leads us to ask: What kind of anthropomorphism is at play in the idea of the Mother Tree and its popular scientific reception? Is this an anthropomorphism that we can accept without demur? What is the relationship between this anthropomorphism and the animal model of sex that dominates in discussions of plants? In this chapter we address these questions via Eduardo Viveiros de Castro's work on what he calls the Amerindian concepts of 'perspectivism' and 'multinaturalism'. Viveiros de Castro's work allows us to make a distinction between traditional and perspectivist anthropomorphisms, where the mark of the latter, in contrast to the former, is that it does not take the nature or meaning of the human for granted. Viveiros de Castro's comparative method also offers a way of thinking the relationship between indigenous and Western thought that avoids some of the aporias that Wall Kimmerer and Turner identify, to the extent that it is thought as transformative translation – specifically, a transformation in the Western conceptual apparatus from the perspective of which the comparison proceeds. Chapter 6 ends with the suggestion that this method of transformative comparison might offer a model for the comparison between plants and animals, avoiding both the recourse to the zoological model that has characterized the history of such comparisons and the epistemological dead end of the idea of the absolute alterity of plants.

The book ends with a reflective Epilogue, in which we try to understand, given our problematization of plant sex, the precise nature of the generic concept of sex. If, as this book argues, the allegedly generic concept of sex in fact presupposes an animal model, and if this model is inadequate to plant sex, how can we continue to use the generic concept of sex to refer to a phenomenon common to different realms of life? What is the nature or form of the commonality of 'sex' in plants and animals, if such commonality there is? And what are the implications of this for thinking about our own, human sexes? Is it possible that the only generic concept of sex is in fact 'vegetal'?

*

One of the major challenges we faced in writing this book was the marshalling of large bodies of literature, from different intellectual epochs and from what are now thought of as different disciplines and/or genres (notably philosophy, history of natural history, history of botany, cultural history, the modern plant sciences, popular science and anthropology). Inevitably, there is a lot that has been left out that is no doubt of relevance, including nineteenth-century German philosophy of nature and all manner of twentieth- and twenty-first-century philosophies of difference. We have drawn very little from science and technology studies (STS). We have not included studies of sexual exuberance or hermaphroditism in the animal kingdom, the phenomenon of 'sex change' in plants or consideration of the detail of the extraordinarily complex life cycles and reproductive systems of algae or fungi. We hope, nevertheless, that what we have done is enough to prove our major point: that there is nothing obvious about plant sex, and that the difficulty of saying what we mean by 'male' and 'female' in plants speaks to a larger problem with the generic concept of sex, which does not leave us – humans – untouched.

1 WHAT IS PLANT PHILOSOPHY? WHAT IS IT NOT?

n the Introduction we distinguished between plant philosophy and popular plant and fungal advocacy. These share the aim of bringing the often startling and unexpected complexity of plant life to greater attention, but otherwise soon part ways. In this chapter we will outline the general shape of these fields and argue that it is only in plant philosophy that we find a concerted attempt to correct the zoocentric and anthropomorphic approach to plant life, even though both fields claim to do this.

It is important to establish what is specific to plant philosophy, particularly the anti-zoocentrism of its approach, in order that the problem of vegetal sex, which we address in detail in the following chapters, appears in the clearest light. This chapter thus lays the ground for the analyses of the subsequent chapters by explaining the critical, anti-zoocentric approach of plant philosophy (in distinction from plant advocacy) that orients this book's investigation of vegetal sex. In explaining how philosophy, specifically, interests itself in plant life and what this interest in plant life offers both to philosophy in general and to botany, this chapter also stakes the claim that the problem of vegetal sex is a philosophical problem. This means not only that the question of vegetal sex can be and has historically been articulated in philosophical terms but that in a sense it still is, and that it is better, then, that we do so wittingly and critically, rather than with the unthought presuppositions of previous articulations.

As we have said, the plant advocacy genre is a popular presentation of recent innovations in scientific research on plant behaviour. This chapter will focus on two broad themes within it: plant sensing and plant intelligence. In the first part of the chapter we will look briefly at the research underlying claims about the senses of plants, arguing that the interpretation of the scientific claims – sometimes by those same scientists themselves – actually shores up, rather than corrects, the zoocentrism and anthropomorphism that it claims to counter, because it ends up arguing that plants are more like us than we (already) think.[1] The plant advocacy literature, we will argue, pits itself against the kind of *moral* zoocentrism that assumes animal life to be more interesting, more complex and ultimately superior to plant life. But it does so only by falling prey to another kind of zoocentrism – the presupposition of *animal models* in the understanding of vegetal life.

In the second part of this chapter we then look at the most influential work on the idea of plant intelligence, that of the biologist Anthony Trewavas. Although Trewavas's work is interpreted anthropomorphically in the plant advocacy literature, we will argue that it is of a different order than the discussions of plant sensing. The idea of plant intelligence, we will argue, is derived from a philosophical proposition – it is not a postulate of empirical science. But when it is taken to entail intention and purposeful behaviour it falls back into precisely the kind of zoocentric model that the idea of plant intelligence has the power to displace.

In the third part we will show how plant philosophy is constituted against this second kind of zoocentrism – against the presumption of the animal model – and how it aims, to the extent that this is possible, to think the plant 'from the plant'. We will show how plant philosophy, unlike plant advocacy, attempts to appreciate the radical alterity of plant life in order to think it – to the extent that this is possible – outside of or beyond the zoocentric models. This is especially true of the botanist-philosophers Francis Hallé and Jacques Tassin. But this does not necessarily mean the reassertion of a traditional distinction between plant and animal (especially human) life. Some plant philosophy (especially that of Michael Marder and Emanuele Coccia) also argues for the recognition of this radical vegetal alterity *within* animal life, including within ourselves. However, having shown the different orientations of plant advocacy and plant philosophy, this chapter ends by identifying their convergence in what is a blind spot for both. In the history of botany, the zoocentric model has left its deepest mark in discussions of plant sex. Contemporary plant science rarely questions the adequacy of the terminology inherited from zoology when it comes to plant sex – although, ironically, this questioning *is* a major part of the philosophical tradition in botany and the history of the theory of plant 'sexuality'. It is striking that plant philosophy, which in all other respects challenges the zoocentric model at every turn, remains surprisingly quiet about plant sex. But can we assume that 'sex' means the same in relation to both animals and plants? That is the central question of this book.

The senses of plants – Are they really 'just like us'?

What is 'plant behaviour'? Plant behaviour, according to Anthony Trewavas, is 'what plants do',[2] rather than what they are, or what attributes they have. Plant behaviours include foraging for light, water and nutrients, which involves the movement of roots (as directional growth) and leaves and reactions of various kinds to stress or attack (e.g. leaf closures and emission of noxious substances, but also growth and gross morphogenesis itself, which is characteristically extremely plastic in plants).[3] Throughout the twentieth century attempts to understand these plant behaviours mostly referred to the fundamental aspects of plant physiology – cell biology, biochemistry and metabolism – and to genetics, evolution and ecology. Effectively, there seemed to be a presupposition that all plant behaviour could in principle be explained in biochemical terms as automatic reaction or reflex.

By the late twentieth century, however, a controversial new tendency in the scientific study of plant behaviour had emerged, insisting that its descriptions and explanations be expanded to include reference to phenomena normally only associated with animal (or even more exclusively, human) behaviour: sensory perception (sight, hearing, taste and so on), communication, intelligence, intention, memory and learning. These new developments (associated especially with Anthony Trewavas, Stefano Mancuso, Frantisek Baluška and Monica Gagliano) do not reject the more established aspects of plant science; its innovations are conceptual and terminological. But it does involve a significant and still controversial shift in the basic conception of plant life.

Claims about the sensory and communicative life of plants have been particularly enthusiastically received in popular discourse. The scientific basis for these claims concerns mostly 'sight' and 'hearing' and communication through 'touch' and volatile chemical signalling. So what does the science say, and how is this popularly represented?

First, can plants see? Everyone is agreed that photosynthesizing plants have different types of photoreceptors for the detection of different light wavelengths. Via these photoreceptors, plants receive information about the light environment. This information must somehow be connected with other kinds of information (e.g. about temperature or season), either from these same photoreceptors or from biochemical responses.

The coordination of this information regulates plant behaviour.[4] It is uncontroversial, then, that plants detect light in this way. But it has been argued that this photoreceptive capacity in plants is involved in more than just behaviour that, for example, maximizes access to light. Photoreceptivity has been linked to 'kin recognition' in plants, where 'recognition' is marked by leaf growth in plants away from neighbouring kin plants to allow them access to light. This occurs, it is proposed, because plants generate specific patterns in reflected light signals that are recognized by their kin. The experiments leading to these hypotheses included screening light from kin plants grown together and varying 'normal' plants with mutants lacking specific photoreceptors. The authors of this study also suggest that the action of the plant hormone auxin was involved in this 'kin recognition', as mutants lacking the ability to synthesize this hormone were not 'recognized' as kin.[5]

All of these claims can be, and have been, made without any reference to the capacity for sight. But František Baluška and Stefano Mancuso refer to the research on kin recognition and to the phenomenon of leaf mimicry described in a paper by Ernesto Gianoli and Fernando Carrasco-Urra to argue that 'higher plants experience a sort of vision using plant-specific ocelli' – 'little eyes': 'the upper epidermal cells of many leaves [that] are shaped like convex or planoconvex lenses capable of bringing convergence of the light rays on the light-sensitive subepidermal cells'.[6] According to Baluška and Mancuso, the research on kin recognition and leaf mimicry suggests that there is some sort of 'plant-specific vision (perception of body shapes of neighbouring plants) via a specific sensory system capable not only of sensing but also of decoding projected images'.[7]

Baluška and Mancuso's suggestion is highly controversial. In a response to their interpretation of his research Gianoli wrote that 'the idea of vision in higher plants not only is unsupported by the facts, but also, and more importantly, is not necessary'.[8] But in the popular communication of plant science even the report of the existence of photoreceptors or the mere fact of phototropism is quickly translated into the idea that plants can 'see'. According to Fleur Daugey, for example, Charles and Francis Darwin's experiments on phototropism mean that 'we can agree that plants possess the sense of sight because they perceive light'.[9] She says that Crepy and Casal's research on the role of photoreceptors in kin recognition means that plants 'see each other', even though Crepy and Casal say no such thing. In his popular works Mancuso similarly equates perception of light with sight to argue that plants do indeed have this sense – indeed 'plants have all five senses, just like us'.[10] The forester and popular science writer Peter Wohlleben infers from the relation between length of day and the triggering of spring-time growth that 'trees must have some kind of ability to see'.[11] The plant geneticist Daniel Chamovitz is torn between an analytical recognition of the fundamental difference between animal sight and plant photo-receptivity and the affirmation that 'plants and humans "see" in essentially the same way'. He writes that the Darwins' experiments proved that it is the shoot tip of a plant that 'sees the light', and that plants have a 'visual system' although of course 'they don't see in pictures. Plants don't have a nervous system that translates light signals into pictures, instead, they translate light signals into different cues for growth. Plants don't have eyes, just as we don't have leaves. But we can both detect light'.[12] Yet detecting light is not the same as seeing.

Similar kinds of claims are made about plant 'hearing'. Vibrations are a constant feature of the subterranean environment, and it is not much of a stretch to presume that living beings rooted in the ground might be sensitive to these vibrations.[13] Plants are known to be hydrotropic – that is, their roots will grow towards soil with a higher moisture gradient. In experiments conducted by Monica Gagliano and Martial Depczynski, seedlings were planted in a tub that divided below into two broad pipes, one pipe resting on a water tray and one not; perhaps unsurprisingly, given what is already known about hydrotropism, the seedling roots grew down into the pipe resting on the water tray. When the experiment was repeated with tubes of running water wrapped around one of the pipes (so the plants could not access the water) the seedling roots again grew towards this pipe. But when this was replaced with a recorded sound of running water the seedlings instead grew away from it. Gagliano and Depczynski interpret these results to show that the seedlings 'respond to acoustic vibrations generated by water moving inside pipes and propagated through the substrate'; that is, that plants 'detect sounds' and indeed 'discriminate sound sources'.[14]

In this paper Gagliano and Depczynski do not claim that plants 'hear', but the equation of 'vibration' and 'sound' or the characterization of the vibrations as 'acoustic' implies an aural mode of sensitivity to vibration. Daugey's popular presentation of this research then represents it under the headings of 'hearing' and 'listening roots'; Mancuso similarly ascribes to plants the ability to hear.[15] In their justifications of this both note that animal audition is reception of vibrations as sound waves. But for the reception of vibration to be experienced as 'sound' requires a specific receptive apparatus and, as Jacques Tassin writes, there is no evidence for this in plants.[16] One may (and scientists do) speculate as to the mechanoperceptive means of reaction to vibration in plants,[17] but to describe the phenomena in terms of acoustics and sound is already a strong interpretative gesture and determines its popular presentation as 'hearing'.[18]

Some of the interpretation of the fact that plants generate sound is more controversial still. When air bubbles form in the system of vessels within the xylem of a tree, rupturing the water column, measurable sounds are emitted. This is called 'cavitation'.[19] Other ultrasonic emissions are also measurable, especially in drought-stressed trees.[20] Gagliano, Mancuso and Robert also present evidence for 'structured acoustic emissions in the form of loud... and frequent clicks' from the young roots of corn.[21] Although the authors of this paper discuss aspects of the interpretation of these sounds in tentative terms ('In plants both emission and detection of sound may be adaptive'), they also discuss the possible futures of such research in such a way as to reveal their belief that these emissions are in fact forms of communication.[22]

Other areas of plant science are also controversially interpreted in terms of plant communication. For example, it is clearly established that plants release volatile organic compounds in response to various stimuli and that these volatile substances are received as signals that provoke further responses from different parts of the same plant, from other plants and from animals, insects especially. These substances may be airborne. The best-known example is the role of the release of ethylene gas from fruits in promoting ripening in other, nearby fruits. Some plants respond to herbivore damage to leaves by producing molecules – in the damaged leaf and in other leaves in the same plant – that deter the herbivore. The fact that some other, undamaged plants in the vicinity of the damaged plant may also produce those substances strongly suggests that signals pass from plant to plant, not just in the air but also through underground mycorrhizal networks linking plants via their roots.[23] It seems that some of these emissions are likely 'a passive consequence of damage to the compartments' in which the compounds are stored (as when cut onions release compounds that irritate human eyes) and others the result of 'de novo synthesis [which] is tightly controlled'.[24] Plants under attack by herbivores may also emit substances that attract the predators of those herbivores.[25]

It might perhaps be less controversial to call this plant behaviour 'signalling', but if 'communication' is taken to mean the exchange of information 'regardless of "intent" or fitness consequences for either party',[26] perhaps few would object to the use of the word 'communication' in relation to plants. But the popular scientific interpretation of these phenomena is strikingly anthropomorphic – a situation for which the scientists have to take some responsibility.[27] Volatile signalling is represented in the plant advocacy literature and popular science more generally as plants (especially trees) talking to each other, using their chemical signals as 'words' to warn each other of potential threat. Because plants stand to benefit from being warned of potential predators, those that 'warn' are understood to be behaving altruistically; plants are said to 'talk' with and act altruistically towards their kin especially.[28]

This anthropomorphic translation of plant science in popular forms (but also, often, in the titles of scientific papers) is of course itself a communicative device; it attempts to present the unfamiliar in familiar terms as an aid to public understanding. But in so doing the specificity of plant life is neglected and the possibility that study of it might give rise to ideas *the very terms of which are unfamiliar to us* is foreclosed. When the sundew leaf bends, over the course of a few hours, to enfold a fly caught on its sticky hairs, we can say that it has moved like a slow hand, but more interesting is the capacity for rapid differential growth of the cells on one side that allows for this bending. What is specific to the plant is that this movement *is* growth,[29] its behaviour *is* morphological development, and it demonstrates what it does not share with animals: phenotypic plasticity. Plant photoreceptors can of course be compared to animal photoreceptors, but the specificity of the behaviours and processes dependent on the different kinds of plant photoreceptors – photosynthesis, phototropism, many more forms of photoperiodism than are found in animals – are ill-understood by being subsumed under the zoological category of sight. The capacity of plants to sense and react to vibrations, not just in one part but differentially across the whole organism, indicates forms of sense that exceed hearing. And volatile chemical signalling and mycorrhizal communication is really very little like talking. To force the description of these forms of communication into the model of speech is lazy and overlooks what is specific to it: the forms of reception of the volatile signals (which are certainly nothing like hearing), the capacity for rapid synthesis and release of chemical volatiles, the range of chemical signalling across the plant world and the nature and complexity of the mycorrhizal relations that facilitate subterranean communication.

It perhaps makes more sense to compare the reception of volatile signals to smell, as Chamovitz does, although the claim that 'plants… undoubtedly smell other plants' is actually open to doubt.[30] But anyway, why limit the possibilities for understanding plant receptors in this way? According to Mancuso, 'plants have all five senses, just like us. And… fifteen others'.[31] But he devotes thirty pages of his *Brilliant Green* to plant sight, smell, taste and touch and only three pages to the other fifteen senses, only three of which he identifies: sensitivity to water (reaction to the humidity in soil and hydrotropism), to gravity and to electromagnetic fields. When, then, it is becoming known that plants have sensitive capacities that either exceed in subtlety or are simply quite other than the traditional five human senses, why would we continue to approach plants only through what we know of ourselves? The plant advocacy literature often claims to want to speak from the plant's point of view,[32] but more often does quite the opposite by anthropomorphizing plant life.

Without exception, the plant advocacy of the past couple of decades wants to distance itself from the Ur-text of the genre, Peter Tompkins and Christopher Bird's *The Secret Life of Plants* (1973).[33] Tompkins and Bird claim, amongst other things (on the basis of reports of mostly extremely dubious and unscientific experiments), that plants can read human minds (e.g. can pick up on guilt feelings in people who have previously harmed plants), across hundreds of miles, and also that they pick up signals from aliens. Although the book reads today more like *Tristram Shandy* than a scientific text (so it is an enjoyable read), it was for a while in the 1970s extraordinarily popular and scientists felt that they needed to rebuff its claims, rather than ignore it. Scientists in general ignore it now, but plant advocates and scientists like Mancuso are still haunted by it. Chamovitz, for example, begins his *What a Plant Knows* by distinguishing himself from Tompkins and Bird: 'My book is not *The Secret Life of Plants*; if you're looking for an argument that plants are just like us you won't find it here.'[34] But in identifying plant sight, hearing, smell and touch the literature often does, in fact, argue that plants are 'just like us': 'plants have all five senses, just like us';[35] 'Leaves, roots and stems are covered with olfactory receptors, like so many invisible noses.'[36] Almost every chapter title of Wohlleben's *The Hidden Life of Trees* is a gross anthropomorphic projection, as if it were not possible to think trees except on a human model ('Social Security', 'Forest Etiquette', 'Tree School', 'Community Housing Projects', 'Street Kids' and so on).[37]

This anthropomorphic tendency in the popular literature dominates in popular discussions of plant sex, where authors struggle to resist lewd (not to say juvenile) metaphors and often remain oblivious to the crude sexual stereotyping involved in the identification of flower characteristics with female human sexuality, especially. Ruth Kassinger's *A Garden of Marvels* includes a chapter called 'Cheap Sex', where the bee orchid's visual mimicry of the female bee and its odour is 'just a sexy skirt and the right perfume'.[38] Michael Allaby's *Plant Love* includes chapters titled 'Red Light Districts' and 'Tarts and Hookers': 'Meet the flowers that are wide open and available to all insect life. These plants are not in the least fussy about who tramps across them – they are the trollops of the hedge and herb garden'.[39] Sharman Apt Russell's *Anatomy of a Rose* includes a description of various garden plants in terms of human sexual and desiring behaviours and is also characteristic in pretending to be embarrassed about plant sex.[40] Perhaps it will be said that this kind of anthropomorphic personification in the popular literature is as irrelevant to the contemporary plant sciences as the anthropomorphic animals of Disney films are to contemporary zoology. However, as we shall see, the condition of possibility of this anthropomorphism – the conflation of sex difference with sexed individuals and sexual parts, which the earliest plant philosophy was careful to distinguish – is also characteristic of discussions of plants in the contemporary plant sciences, which thus share a metaphorical space with this popular literature, albeit its manifestation is less egregious.

Plant intelligence – but not 'intention'

Some of the scientists who object to the idea of plant sight and hearing and so on do so because they think that it is unhelpful and scientifically illegitimate. For the same reasons some also object to the idea of plant intelligence. But the question of plant intelligence is actually of a different order. The claims about plant 'sight' and so on are interpretations of the results of experiments, and they ascribe specific capacities to plants or to parts of plants according to analogies with animals. This analogical model has been a feature of Western investigations of plants since at least Greek antiquity, as we will see in Chapter 2. The question of intelligence, on the other hand, is from the beginning far broader and concerns the conceptualization of life itself, although this passes without notice in the plant advocacy literature. The question of plant intelligence arises at the point of indifference between plant science and philosophy. It is not a matter of inferring the possession of a faculty on the basis of experimental data, no matter how fine-grained or acute that data might be. Plant intelligence is not something that is posited as a result of an investigation into specific living beings. The idea of plant intelligence – indeed of intelligence itself in biology more generally – frames any investigation of living beings. It is a regulative idea that orients the investigation of plants.

The idea of plant intelligence is most closely associated with Anthony Trewavas, whose work is very frequently cited in the plant advocacy literature. However, these references tend to be reduced to various definitions of intelligence, according to which plants can be called intelligent, while Trewavas's extensive work on the topic suggests much more than matters of definition.[41] True, Trewavas himself often gives and discusses definitions. He borrows from David Stenhouse the very general idea that intelligence is 'adaptively variable behaviour within the lifetime of the individual'. As plant behaviour (response to internal and external signals) is typically expressed as growth and development, plant intelligence may be defined as '*adaptively variable growth and development during the lifetime of the individual*'.[42] Adaptive behaviour is a response by the organism to a problem presented by the environment, which also gives rise to the definition of intelligence as a capacity for problem solving.[43]

According to Trewavas and others, the major obstacle in accepting the idea of plant intelligence is anthropocentrism, and more specifically the conflation of intelligence with human-style cognition, or the presumption that intelligence is equivalent to the intellectual capacity for 'cleverness'. With this anthropocentric prejudice, plant behaviour is reduced to forced reactions or response reflexes, on the outdated model of animal behaviour proposed by Jacques Loeb in the early-twentieth century.[44] Trewavas argues, instead, as others have done, that human-style cognition (and 'cleverness') is only one form of the expression of intelligence. Intelligence in general is adaptive behaviour or problem solving and all living organisms must be intelligent if they are to survive.[45] Organisms must react to specific (and indeed each time unique) sets of circumstances in specific ways and have some way of 'assessing' those circumstances so as to produce the most appropriate response – where the capacity to 'produce the most appropriate response' just is fitness. All living organisms are faced with basic biological problems (finding food, avoiding predators and so on) and the ability to deal with these problems – or the ability to survive – is 'intelligence'. In plants intelligence is 'adaptation of the individual solving its local problems by phenotypic plasticity and if we could monitor it sufficiently well, metabolic and genomic plasticity, too'.[46]

Perhaps surprisingly, Chamovitz objects to Trewavas's extension of the idea of intelligence to plants, but this is because he misinterprets it. He takes the idea of plant intelligence to refer to a biological characteristic 'no different from, say, body shape and respiration' – that is, an evolved characteristic.[47] Chamovitz says that Trewavas recognizes that humans are more intelligent than other animals and plants, but the idea that we could compare the intelligence of difference species quantitatively does not in fact make sense given Trewavas's definition. Trewavas continually stresses the relationship between intelligence and fitness. Intelligent behaviour can only be expressed within the confines of the behavioural capacities of an organism,[48] and of course those capacities vary from organism to organism. If intelligent behaviour maximizes the fitness of an organism, each different kind of organism will have its own set of intelligent behaviours and it makes no sense to say that an organism that does not have the human set of behaviours is less intelligent than the human, because it has the right set of intelligent behaviours to ensure its own survival and maximize its fitness.

But more importantly, intelligence is not one capacity of an organism amongst others. When Trewavas develops his argument at length in *Plant Behaviour and Intelligence*, it becomes clear that he approaches the issue from a systems theoretical approach. He describes a system as 'a network of mutually dependent and, thus, interconnected components comprising a unified whole'.[49] Living organisms, considered as individuals, are (biological) self-organizing systems, although the environment is arguably also part of each. Species may also be considered as systems, as may the earth and its atmosphere, the solar system and so on. Swarms, hive colonies and colonial organisms such as coral are systems. At the microscopic end of the scale, each cell is a system, and within the cell there are yet smaller subsystems, for example, 'aggregates of enzyme sub-units to form a fully functional enzyme'.[50] All systems possess emergent properties 'that is, properties that, to all intents and purposes, are not shared by the individual constituents, but only by the system as a whole'.[51] Intelligence is one such emergent property, but it does not emerge as one property amongst others of equivalent importance, much less evolve. Effectively – although he does not ever articulate this conclusion explicitly – for Trevavas *biological intelligence is life itself*. To be alive is to be a biologically intelligent system, or an aggregate of such systems; to be alive is to be a biological system that adapts to the problems posed to it by its environment and responds successfully enough to maintain the system. There can be no non-intelligent life.[52] Intelligence is not a property that a biological system may or may not have, and the problem is not whether we can infer the presence of intelligence from manifest behaviours or whether an organism meets the criteria, however they are defined, for intelligence.[53] The postulation of plant intelligence in Trevavas's work is not the result of a series of investigations; it is ultimately a kind of philosophical axiom that orients research.[54] It is not that the study of plant behaviours leads one to believe that plants are intelligent; plant intelligence is an axiomatic postulate that allows one to study plant behaviours as adaptive, allows one to understand adaptively variable growth and development as such.[55]

Without endorsing this axiom as such but noting its place within the system of knowledge that develops in Trevavas's work, we can say that within this system, and with this axiom, the idea of plant intelligence is perfectly acceptable and not only is it not anthropomorphic, it is also explicitly anti-anthropocentric because it denies that intelligence is a uniquely human trait. But once we have accepted the idea of plant intelligence, do we also have to accept plant learning and plant memory? Do we also have to accept that plants act intentionally and with purpose, as Trevavas claims that they do?

Trevavas points out that informational models allow us to speak of 'learning' and 'memory' without necessarily implying cognitive processes.[56] As with machine memory and learning, it is possible to use these terms without implying any kind of reflective, intra-psychic process. Evidence for memory and learning (the two are inseparable) in plants concerns either specific plants and their specific actions (often, the Venus flytrap) or more general behavioural phenomena, especially concerning light and temperature sensitivity. As is now well known, thanks to the popularizing efforts of David Attenborough in particular, the Venus flytrap snaps shut when its prey touches not just one but two of the sensitive hairs on its lobes within about twenty seconds.[57] This leads several influential plant advocates (including Attenborough) to say that plants can count or remember how many of its hairs have been triggered or that it possesses a short-term memory.[58] The Venus flytrap's behaviour has been explained biochemically in terms of the accumulation, to triggering intensities, of bioactive metabolites. With this explanation the description of the behaviour as involving a 'memory' phenomenon (the plant 'remembers' that and when the first hair was touched) is only metaphorical; indeed, the explanation does not require the use of the idea of memory at all.[59] Those happy to speak more systematically of plant memory and learning in respect of general phenomena may propose physio-chemical explanations which, in the simplest terms, involve the traces of received signals and experienced events (including seasonal changes in temperature or herbivore attacks) being integrated into and thus modifying the dynamic state of the living system.[60] For example, a plant that once experiences insect attack and responds appropriately may then be 'primed' to respond more quickly in the event of a future attack, even if that attack is not in fact forthcoming. A systems theoretical approach allows one to call this 'memory' and 'learning' without unduly anthropomorphizing the phenomena – although its presentation in popular plant advocacy is thoroughly anthropomorphic, once again.[61]

But the same defence cannot be made for the claims that plants act intentionally and with purpose. Trewavas's discussion of purpose and intention does not take his position on plant intelligence to its logical conclusion; rather, it betrays it. This aspect of discussions of plant intelligence and behaviour seems to be driven more than anything else by a misplaced desire to claim every honour for plants that is awarded to animals. In Trewavas, where we find the longest discussion, the claims tend to be made, moreover, on the basis of a confusion over the different meanings of 'purpose' and its effective conflation with 'intention'.

Trewavas claims, no doubt correctly, that no plant that behaved or grew 'randomly' would survive. Even growth that might seem random – for example, a root structure that at a certain moment, when pulled from the ground, might look to have sent out its root hairs randomly – is rather the effect of searching behaviour, looking for signals (moisture indications) in their absence.[62] Within the genetically determined constraints of the organism, this behaviour can be called intelligent – variable growth and development in response to a problem posed by the environment (here, lack of water).

But Trewavas also wants to call this behaviour 'purposeful', as part of an effort to reintroduce teleology (under the name of 'teleonomy') into biology. At one point he defines teleonomy as 'the apparent purposefulness and the obvious goal-directed nature of living organisms that derives from both adaptation and evolution.... I regard [teleonomy] as the obvious drive to survive and reproduce.'[63] Later 'purpose' and 'goal' are used interchangeably, where it is said that the purpose or goal of intelligent behaviours is survival and 'the attempted optimization of fitness'.[64] No doubt intelligent behaviours can be said to be 'purposeful' to the extent that they are adaptive or functional, and some do argue that a teleological viewpoint, in Immanuel Kant's sense of an orientation of thought, is an ineliminable part of biological understanding.[65] But Trewavas confuses the overarching idea of functional purposefulness in the behaviours of living beings with their having and *intending* a purpose in each behaviour. Misled by a dictionary definition of 'intention' as 'purpose or goal', he constructs a specious argument: if 'intention' means 'purpose' or 'goal', as his dictionary tells him that it does, and if the purpose or goal of an intelligent behaviour like gravitropism, for example, is survival, then the plant *intends* to react to gravity.[66]

Trewavas's position develops over the years across a number of different texts and the terminology is not always used quite consistently at each point. But the general movement of his thought on the issue of plant intelligence and intention can be reconstructed in such a way as to allow us to see the emergence of a hidden, paralogistic form of reasoning. Distilled in the way that Immanuel Kant distilled a whole tradition in the first paralogism of pure reason (concerning the substantiality of the soul), we see a term – 'intelligence' – that shifts in its meaning across the major and minor premises of an argument to give an unwarranted conclusion.[67] In the major premiss biological intelligence is life itself. The construction of this premiss – a significant 'emergent' philosophical achievement of Trewavas's *Plant Intelligence and Behaviour* – is more than a hypothesis within a system of biological knowledge; it is the orienting point, or something like a transcendental epistemological condition, of that system of knowledge itself. To accept it involves a shift of perspective that decentres the human, who no longer stands at the centre of creation as *the* intelligent animal, pleased to find itself reflected in 'lower' forms of life. The second premiss, that plants are living and therefore intelligent, is in fact contained in this first. But in being presented as a separate, subsidiary premiss, 'intelligence' becomes one term in a constellation with others – purpose, intention, cognition, consciousness, decision, choice – and plays its part as a constituent (not an orienting point or transcendental condition) in a series of claims. If the second premiss is 'plants, as living beings, are intelligent', the conclusion 'plants therefore have intentions and purposes' only follows if 'intelligence' is a specific kind of property that brings these things along with it. But this is to return to the anthropo- or zoocentric perspective which the first premiss displaces. The radical anti-anthropocentrism of the shift in perspective effected by the major premiss cannot be sustained after the unconscious slippage into a more traditional framing of intelligence in the second.

The regression to an anthropomorphic frame brings with it a vocabulary of 'decision' and 'choice', and it is especially jarring when Trewavas and others speak of the *evolutionary* choices and decisions that plants have made. Modular development, and the capacity to regenerate various different parts of the plants from unspecialized cells, is said to be a strategy that is 'the result of an active evolutionary decision to place survival above specialization'. Plants, unlike animals, chose 'an outer, relatively rigid [cell] wall'.[68] Mancuso, too, often speaks in this way: plants, according to him, made 'a choice in favour of a stationary life' – this was their 'initial choice', 'the primitive choice to evolve as beings that are stationary'; 'as an evolutionary strategy, plants... have chosen to be composed of divisible parts' (modularity was 'a fundamental strategic choice'); plants choose which animals to enlist as helpers in pollination.[69]

These are not just metaphors. They are the symptom of a regressive tendency in the plant advocacy literature and in some of the science on which it is based. This stems from the fact that its initial anti-anthropomorphic impulse concerns not resistance to animal or human models in the investigation of plant life but resistance to the denegation of plant life and to the presumption of the superiority of animal, and especially human, life. In countering this moral anthropomorphism, the plant advocacy literature presupposes that if we are unwilling to recognize intentional behaviours, for example, in plants, it is only because we value ourselves too highly and cannot bear to share our animal or specifically human characteristics with plants. It presupposes that to recognize complexity or biological sophistication in plant life, we must be able to show that it is, in fact, intrinsically similar to the complexities of animal life. On this view to deny plants cognition or consciousness is a priori to denigrate them. Yet this view resolves itself into another – and more theoretically significant – form of zoocentrism and anthropomorphism, because it has continual recourse to animal or specifically human models in the defence of plants, as if appreciation of the complexity of plant life rested on it being able to be shown that plants see, hear, smell, taste, touch, think, are rational, process information cognitively, are conscious, make decisions and assess future returns on the investment of energy. But what if the specificity of plant life – in all its complexity and biological sophistication – was located somewhere else entirely? What if the continued attempts to favourably compare plant and animal life shielded what is specifically *plant* from our animal view?

Beyond the zoocentric model: Plant philosophy

This is where plant philosophy steps in, with its aim to think plant life outside of zoocentric models by focusing on what is specific and peculiar to plant life in relation to the animal.[70] All plant philosophy sees, in one way or another, that this task of understanding plant life cannot be a purely biological or botanical task, but is one which requires some philosophical thinking. They all see that, to a greater or lesser extent, and in different ways, philosophy and botany are entwined. Plant philosophy reveals, moreover, the zoocentric form and presuppositions of many of the traditional categories and conceptual oppositions that continue to structure much of Western philosophical thought. 'Western philosophical thought' is itself a problematic category. The geopolitical idea of the 'West' that it encompasses is a younger invention than that of the 'history of philosophy', to which it also refers. But since Martin Heidegger's attempted 'destruction' of metaphysics, 'Western metaphysics' has come to name not so much an empirical or textual history as a problem. For Heidegger this was the problem of the 'forgetting of Being', but for various post-Heideggerian philosophies the problem could be thought otherwise, especially after Jacques Derrida's deconstruction of specific conceptual features of the history of Western thought inspired a generalized deconstruction of Western metaphysics.[71] This is the terrain within and against which existing plant philosophy works. It aims to think plant life outside of the zoocentric models which, it is argued, have hitherto dominated in popular and philosophical views of plants. In so doing, it demonstrates the extent to which the traditional categories of Western metaphysics are inadequate to the task. As the traditional categories of Western metaphysics are supposed to be able to encompass all of reality, their inadequacy to the ontology of the vast majority of the earth's biomass is more than a regional oversight. As Marder puts it, 'in

its very being the plant accomplishes a lived destruction of metaphysics';
they are the 'weeds of metaphysics',[72] growing in, and widening, its cracks.

We could also say that plant philosophy grows in the cracks of botany. In contesting the zoocentrism of the botanical tradition, both the botanist Frances Hallé and the ecologist Jacques Tassin move easily between scientific and philosophical reflection on plant life and similarly contest the adequacy of various metaphysical categories to the ontology of plant life. Tassin casts the central question of his *À quoi pensent les plantes* (*What Do Plants Think?*) in explicitly philosophical terms – not 'what is this plant', or 'what is a plant' but 'what is it to be a plant?'[73] He suggests that many questions in botany lead inexorably to problems that are simultaneously biological and philosophical. A whole new approach to plant life is required, he writes, and plant science alone cannot achieve this:

> There can be no true, intimate, ontological knowledge of the vegetal without an alliance between science and philosophy, and also poetry – all three of these are necessary to bring us closer to the vegetal and, in so doing, to distance ourselves from ourselves.[74]

To this we must add that botany and philosophy have in fact always been entwined; the task now is to make that relationship explicit and thus release botany from its unthinking recourse to traditional metaphysics.

So, what is it to be a plant? What is it in plant life that sets plants apart from animals, that resists zoocentric models and problematizes the traditional categories of Western metaphysical thought?

The principle of organization of a plant body is so different from that of an animal that, in thinking this through, the cryptic metaphysics in which the modern idea of the 'organism' is suspended becomes visible. In the most general biological sense an organism is a synonym for a living entity; in its conceptual genealogy it is an organized body, a self-organizing, integrated whole-part unit.[75] Animals are (in general – there are exceptions) organisms in this sense: integrated whole-part units, with specific organs for specific functions (albeit they interact), usually under the control of a centralizing (precisely, organizing) function (the brain). Even brainless animals are integrated, with some specific organs for specific functions. Plants, on the contrary, are not centrally organized; they have no brain equivalent. Their functions are not distributed across specific organs; indeed, as almost all plant philosophers and advocates point out, plants have no organs.[76] The principle of organization of a plant is modularity; that is, plants are made up of repeated 'units' (although, as we shall see, even this word is questionable). Each shoot (so, on a tree, each branchlet or twig) produces, indefinitely, reiteratively, functional units (called phytomers) comprising bud and leaf and often flower. In this way the plant clones itself – clones its basic form – on itself, indefinitely. It has no final form.[77]

A modular principle of organization means that damaged or lost parts can be replaced, which is essential for beings subject to herbivory. It means that a plant can be cut up into separate parts, which can then each continue to grow independently. This regeneration and reiteration of parts is possible because plant cells are not specialized, even though they become differentiated. In the development of an animal, embryogenesis is determined; that is, so-called fate determination of cells occurs quite quickly – cells differentiate in order to perform specific functions (to be part of specific organs) and it is only under certain relatively rare conditions that they can de-differentiate and perform a different function. Only the original animal zygote is totipotent; that is, only the zygote has the capacity to divide and differentiate into all of the different types of cell in an organism.[78] Once differentiated, a liver cell or heart cell, for example, will not, under normal circumstances, become a cell of another kind.[79] In plants, on the contrary, many cells are potentially totipotent. As budding from the plant stem shows, the capacity for cell de-differentiation (or reversibility of differentiation) is the rule, not the exception. A bare cutting from the stem of a flowering plant, a section with no buds or leaves, planted in soil can grow roots and buds and flowers – the entire plant form can be reconstructed relatively quickly from only one part of it.[80] The embryonic cells at the tips of roots and stems (meristems) will anyway continue to produce the cells needed for all parts of the plant. For Hallé this justifies us in saying that the plant's embryogenesis is indefinite, not limited to a short period at the beginning of its life, like an animal.[81] Because the form of the adult plant is not visible in the 'embryo' of the plant, Tassin interprets the situation somewhat differently: the indefinite growth of the plant means that its development is not embryonic but 'postembryonic'. Morphogenesis continues throughout the life of the plant or to put it better, he says, morphogenesis 'begins when embryogenesis ends'[82] – which is again quite different to the animal.

Non-centralized organization, modularity and cell totipotency mean that plants are morphologically extremely plastic. An animal, as Hallé puts it, is essentially a mobile volume with a modest surface area. As it grows its volume increases while its form stays more or less the same. A plant, by contrast, is essentially a surface; as it grows it changes form to maximize this surface.[83] Of course, environmental factors influence the growth and development of animals, which are to some small extent therefore morphologically plastic. Also, there exist spectacular examples of animal metamorphosis. But in both cases (general, limited morphological plasticity in animals and metamorphosis), as Hallé argues, these changes are circumscribed, limited and to a great extent predictable.[84] The change from caterpillar to pupa to butterfly is certainly dramatic, but at each stage the assumed form is predictable and only minimally plastic. The morphological plasticity of plants is not, of course, unlimited, but it is extreme. Trees provide us with good examples. There is an architecture of tree growth,[85] but the modular details are unpredictable and vary considerably from plant to plant of the same species; in this sense the adult form in general is unpredictable. But in the life of a plant there may also be repeated transformations. An upright tree may fall, its trunk now lying on the earth. The trunk may then sprout roots on the earth side and shoots on the aerial side to transform itself into a line of trees. A mature tree with a central trunk that is felled to leave only a stump may then grow multiple trunks from that stump or stool.[86]

Modularity, vegetative reproduction, morphological plasticity and the existence of embryonic stem cells at multiple points in the plant problematize some of the basic distinctions within the animal life cycle, distinctions which a zoocentric perspective tends to think of as basic to the life process itself. The plant is always, for example, an amalgam of parts, some of which are somatically old and some somatically young. The plant is essentially incomplete. Its somatic being is, as Tassin says, perpetually becoming and in a sense it remains perpetually young; or the plant is the site of a 'marvellous dialectic of age and youth'.[87] Many plants, and most trees, grow without any genetically pre-programmed senescence.[88] An ancient tree – counting in thousands of years, not just hundreds – will sprout growth as vigorous, healthy and fertile as a sapling. And the vegetal principle of indeterminate growth confers on it the potential for somatic immortality. Further, as a form of life based on a modular principle, a plant, and a tree especially, does not respect the absolute zoocentric distinction between life (being alive) and death. It may be that it only dies in a succession of deaths; it will often comprise parts that are dead even as it thrives.[89] And it will live out of its own dying, as a tree nourishes itself from its own rotting matter, with the help of fungi.[90]

If plants and animals do not die in the same way, this is also because – Heidegger notwithstanding – the animal death is an individual death, but the plant is, as Hallé puts it, hardly compatible with the idea of the individual.[91] What is an individual? he asks, insisting that this kind of philosophical question, although mostly avoided by scientists, is not irrelevant to contemporary scientific practice in biology. Is there a category of the individual that is appropriate to all of the different forms of life, or is the idea of the individual which we nevertheless apply to all forms of life, including plants and bacteria, in fact a zoocentric, or even more specifically anthropomorphic, idea tied to a metaphysics of identity? Hallé reprises a series of different definitions of the living individual: etymological, genetic and immunological. Etymologically an individual is a living being that cannot be divided without dying. Genetically, a living individual is defined in terms of its having a stable genome both spatially (the same at all points in the individual) and temporally (across the life time of the organism). Immunologically, the living individual is a singular, clearly delimited functional entity differentiated from others and possessed of a 'self' in relation to the 'non-self'.[92] Animals, to the extent that they meet one or all of these requirements, can be said to be individuals, but plants may well not meet any of them and the ways in which plant life exceeds the boundaries of these three definitions are related.

The etymological definition is a conceptual definition. The individual is the concept of that which cannot be divided. But plants of course can be divided – they are precisely not indivisible. Division is multiplication for a plant, not death. But it is also not so simple to say that an individual plant divides to produce two or more individuals because it is not clear where we would 'locate' the individuality of a plant, or which part comprises the individual, in a modular being. Is it the whole plant with all of its reiterations or is it the reiteration itself that is the individual? Is the 'individual', in relation to the plant, therefore a recursive concept, referring to a 'nested' reality?[93] Or is the tree better thought as a population of units that remain attached together, as the biologist John L. Harper suggested?[94] In that case the tree would gain its unity from possession of a common root system, but many plants develop adventitious root systems from buds and other parts which may become detached. Roots from different plants (both of the same and of different species) may fuse; most – perhaps 80 per cent – of the plants survive only because of usually symbiotic relation with fungi, called mycorrhizal relations, where the hyphae of fungi grow in or attach to plant roots to effectively become part of the plant – or is it that the plant becomes part of the fungus? These subterranean mycorrhizal networks can stretch across whole forests, connecting huge numbers of plants and fungi sharing nutrients and water and communicating chemical and electrical signals. Where or what is the 'individual' in these forms of life?[95]

Tassin, the most philosophical of scientists, approaches the problem from a slightly different angle. In his book *Penser comme un arbre* (*Thinking Like a Tree*) he does not begin with definitions and find the plant (or more specifically, the tree) recalcitrant to them; he explores the problem beginning with the tree. As the tree clones itself, he says, it deindividualizes itself, and becomes more and more deindividualized over time. Its existence is simultaneously singular and plural; unity and community coexist without either taking precedence over the other. From the zoocentric viewpoint a clone is the production of another, separate individual, but the plant also clones itself upon itself – its 'descendants' are not disconnected from it. According to Tassin, our animal reality renders it difficult for us to conceive of the capacity of the tree to be both one and many, and yet that is vegetal reality. One may be tempted to reply that it is not our animal reality that determines this, but the presumption of the metaphysical category of the individual and its constitutive opposition to the idea of community, or the metaphysical category of unity and its constitutive opposition to the idea of plurality. But the deeper point – albeit it remains hitherto unexplored in plant philosophy – is the relationship between 'our animal reality' and these metaphysical categories. Tassin writes that there is an 'organic inadequation' between the animal and the vegetal. He implies, with this, further, that there is a conceptual adequation between the principle of organization of animal bodies and the principles of metaphysics. Metaphysics is animal metaphysics.

Plants may also defy the genetic definition of the individual, possessed of its unique spatio-temporally stable genome. As Barbara McClintock first demonstrated in the 1940s (though the work was effectively ignored for some decades), plant genomes are relatively plastic.[96] The 'embryonic' cells of the meristem may give rise to genetic mutations during mitotic division, mutations that may then be somatically expressed on that part of the plant. If the relationship between the mutant cells and the pre-mutant cells stabilizes, the same plant will carry, side by side, parts with different genomes.[97] Some young algal forms fuse to form an adult with a genetic mosaic.[98] Plants also defy the immunological definition of the individual, for both biological and conceptual reasons. Plants do not have immune systems, and yet, in some respects, there is a recognition of 'self'/'non-self'. For example, some plants have biological strategies for avoiding self-pollination; some will fuse root hairs from the same taproot but not another. In other respects, the self/other distinction is much less strict than for animal life.[99] Plants of different species will sometimes fuse roots; cross-species grafts are possible and hybrids are common – so common, in fact, that the taxonomy of plants cannot easily be represented in the same phylogenetic form as animal taxonomies.[100] But the conceptual problem concerns the meaning of 'the self' for a plant. Even where the zoocentric gaze can determine plant individuals for itself – a cutting from a pot plant can be planted in another pot and moved away from its origin, and we can say that these are two individuals *for us* – are divided clones from the same plant still parts of the same plant self, as far as the plant itself is concerned? Does it make any sense to say 'as far as the plant *itself* is concerned'?[101] Are divided plants part of the same plant 'self' or are they other, or kin? Is it the case that the plant self is not the same as the plant individual – if indeed, it makes sense to speak of either?[102]

Thus far we have purposefully concentrated on the plant philosophy that comes from botany, because 'plant philosophy' is a not a sub-area of a self-sufficient discipline. But plant philosophy emerges in these thinkers in the service of botanical and ecological understanding. What they really want to do is understand plants as plants. The plant philosophy that emerges from the discipline of philosophy wants to philosophize *beyond plants*. Michael Marder's book *Plant Thinking: A Philosophy of Vegetal Life* is currently the high watermark of this genre. In it, Marder proposes vegetal life as a model for thinking aspects of all life and for philosophy.[103] He proposes a 'vegetal ontology' which is not just the description of the characteristic features of a region of entities. It is – contra Heidegger's restriction of 'world' to Dasein – an exploration of vegetal existentiality: the temporality, spatiality and affirmative, repetitive expression of vegetal existence. But his aim is to remove, not reinforce, what he calls 'the real and ideal barriers humans have erected between themselves and plants'.[104] The central conceptual means of this attempt is the revalued resurrection of the Aristotelian idea of the vegetative soul. The 'soul', Marder writes, has been expelled from modern philosophy – it is no longer a respectable object of philosophical analysis. At the same time, he claims, in 'the age of positivist science', plants are assigned to 'the specialized (ontic) discipline of botany' and philosophy is thought to have nothing useful to say about plants. The idea of the soul and the realm of plants are thus united in 'their exclusion from the purview of respectable philosophical discourses', but this exclusion is the entry point for Marder's plant philosophy and its rethinking of the vegetable soul:[105]

> Contemporary philosophy disengages from these two entities and, in so doing, abandons them, sets them free. Left to their own devices, each transforms the other: the plant confirms the 'truth' of the soul as something, in large part, non-ideal, embodied, mortal, and this-worldly, while the soul, shared with other living entities and construed as the very figure for sharing, corroborates the vivacity of the plant in excess of a reductively conceptual grasp.[106]

The nutritive, reproductive soul refers to that aspect of life that not only assimilates the nourishing other but also becomes other in its progeny.[107] It is the emblem of the non-identical. As a model of or for plant thinking, a thinking which 'does not maintain an identity *as thinking*', it offers an alternative to the transcendental synthesis of the 'I think'.[108] In plant thinking, 'it thinks': 'At the core of the subject, who proclaims: "I think", lies the subjectless vegetal *it thinks*, at once shoring up and destabilizing the thinking of this "I"'.[109] Ultimately, for Marder, this is the life that thinks through the animal (including the human) and the plant, and this embodied 'life' is 'the unacknowledged vegetal background of human knowledge'.[110] The philosophical encounter with plants thus moves us to recognize the vegetal part within ourselves, and also to render human thinking more plant-like, which means to take on board the challenge that plant life poses to traditional philosophical categories: plant thinking starts 'with the explosion of identity'.[111]

Emanuele Coccia, in his book *The Life of Plants: A Metaphysics of Mixture*, also contests Heidegger's denial of 'world' to animals and a fortiori plants. Coccia argues that plant being is 'the most intense, the most radical, and paradigmatic form of being in the world. To interrogate plants means to understand what it means to be in the world'. As plants transform solar energy into their living bodies, bodies upon which all animal life depends, the life of plants is 'a cosmogony in action, the constant genesis of our cosmos'.[112] Being completely exposed to the world, 'in absolute continuity and total communion with the environment' – indeed, plants make the environment – plant being teaches us that being in the world means 'transcendental immersion': 'There is no material distinction between us and the rest of the world.' 'Everything is in everything', as the title of one of Coccia's chapters has it.[113] Coccia thus proposes a vegetal metaphysics with, for him, boundless implications.[114]

The plant philosophy that comes from botany tends to resist this kind of move. For Tassin, for example, the tree is too different from us for us to be able to take it as a model. What is most important is to recognize the 'full alterity' of the tree, 'an alterity that we accept has an unknown and inaccessible part'.[115] To the extent that Tassin mainly refers here to the extraction of social models from plant life, it is difficult not to agree, not least because the social forms attributed to plants in the plant advocacy literature to which he refers are themselves anthropomorphic fantasies. But if Marder's conception of the vegetative soul refers us to an 'unknown and inaccessible part' of ourselves – perhaps something like a material unconscious? – plant philosophy indeed suggests fascinating unexplored possibilities.

But what about plant sex?

Whether plants are thought as too 'other' to provide a model for thinking any aspect of animal life, or whether 'plant thinking' – perhaps via the idea of the vegetative soul – opens a view onto its vegetative aspect, all plant philosophy is united in aiming to challenge the zoocentric model. This is its fundamental difference in orientation from plant advocacy, which often has recourse to zoocentric models in the pursuit of its moral anti-anthropocentrism. It is surprising, then, that plant philosophy has relatively little to say about the idea of plant sex and the nature of the discourses surrounding it. In the history of botany, and still in the plant sciences today, the zoocentric model has arguably left its deepest and most enduring mark in discussions of plant sex. In contemporary plant science the appropriateness of the terminology inherited from zoology seems largely to be taken for granted, although the early history of the theory of plant 'sexuality' is precisely a debate on the meaning and application of this terminology. And whereas the jettisoning of inappropriate animal models and metaphors is generally seen as basic to the inauguration of scientific botany, the situation is quite otherwise with sex, where progress is measured in terms of the acceptance of the literal truth of the animal model for plants.

As we have seen, discussions of plant sex in the popular scientific literature on plant reproduction take every advantage of the shared fact of sexual reproduction in the plant and animal kingdoms to represent plant sex in anthropomorphic terms. The case is obviously different in the scientific literature (apart from the odd jokey article title), but the technical vocabulary of plant reproduction and ecology extends beyond 'male' and 'female' to include also 'mother', 'sister', 'parent', 'offspring' and so on. Is this a literal or a metaphorical biological vocabulary? On the whole plant philosophy accepts this terminology, or at least refrains from any critical investigation of it. But what, if anything, is the effect of such a vocabulary on the scientific and philosophical understanding of plants? Is there any way that the idea of a 'mother', for example, can be anything other than zoocentric? Is there any way that a terminology derived from human social relations can be anything other than anthropomorphic?

Plants alone, of all living forms, have a dibiontic life cycle – that is, two different multicellular stages, also known as alternation of haploid and diploid generations (in one generation the organism has a single set of chromosomes; in the other it has a double set). Briefly (we will return to this in detail in Chapter 5), the forms of each plant with which we are most familiar – what we generally think of as the plant itself – are just one phase of the plant life cycle, most often the sporophyte but sometimes the gametophyte phase or generation. The diploid sporophyte generation includes cells capable of undergoing meiosis, as in animals. In animals, meiosis produces haploid gametes that may then fuse with other haploid gametes in fertilization (also called syngamy) to produce diploid zygotes. An animal gamete cannot live by itself and only in exceptional circumstances can it grow by itself into a new haploid individual.[116] In plants, however, meiosis produces haploid spores. These spores cannot undergo syngamy (precisely because they are not gametes), but do divide mitotically to produce an entirely new, haploid plant – the gametophyte. The haploid gametophyte produces haploid gametes (as the name suggests) by mitosis; these gametes then undergo syngamy to produce a new diploid sporophyte. The sporophyte and the gametophyte that are the two generations of a life cycle do not resemble each other at all (and the gametophyte may develop within the sporophyte). In flowering plants, the male gametophyte grows within the pollen grains; the female gametophyte is the 'embryo sac' that grows within the ovary. There is nothing like this in animal life: the haploid gametophyte is unique to plant life.[117] And although there are two generations in the 'individual' plant life cycle, there are at least three plants – one sporophyte and two gametophytes (one male, one female). The dibiontic life cycle thus further troubles the idea of the individual organism characteristic of the zoocentric model.

Why has this aspect of the specificity of plant life – perhaps *the most specific aspect* – seemed less interesting to plant philosophy?

The investigation of this and other related questions requires a study of the history of the theory of plant sexuality, which is also a history of a much longer lineage of plant philosophy, from Aristotle, thorough Theophrastus, Nicolaus of Damascus, Albertus Magnus, Andrea Cesalpino, Rudolf Jakob Camerarius, Carl Linnaeus and into modern plant science. This history shows that the basic problem against which plant philosophy constitutes itself – the zoocentric model – is nowhere more obvious than in discussions of plant sex, which can be seen as paradigmatic for that zoocentric model. This is what the rest of this book will trace, arguing that the focus on vegetal sex also brings to the fore various other issues with which plant philosophy is concerned and problematizes the generic concept of 'sex' itself. From antiquity the question 'What is plant sex?', or more properly 'What do "male" and "female" mean in relation to plants?' inevitably raised the broader and explicitly philosophical question of what we mean by 'male' and 'female' in general, or the philosophical question 'What *are* "male" and "female"?' This is where we start in the following chapter.

2 PLANT PHILOSOPHY AND PLANT SEX: ARISTOTLE TO ALBERTUS

What is plant sex? Although any botanist will have a ready answer to this question (male and female in plants are distinguished as bearers of gametes of different sizes), the question is, on closer inspection, particularly complex. The same ready answer serves as a definition of 'sex' in the animal kingdom too. More specifically, sex is defined in terms of anisogamy (reproduction via the fusion of two gametes of different sizes). As Lynn Margulis and Dorion Sagan explain:

> the smaller of a pair of gametes is dubbed 'male'. Male sex cells are so designated, however, not only because of their smaller size but because of their propensity to move on their own.... the practical definition of *male* is simply an organism (gamont) that produces moving, relatively small gametes, often in profusion, but the term may be used for one of those gametes (or microgametes) itself. The term *female*, by general definition, usually refers to those gametes which stand still or those gamonts which produce sedentary gametes.[1]

This scientific definition is obviously remote from the popular, everyday understanding of sex as the bare, biological fact of the difference between male and female. This scientific definition would have to be finessed to encompass all of those popularly identified as male and female (given, for example, that women who have achieved menopause, pre-pubescent boys and 'sterile' adults of all kinds do not, in fact, produce gametes). The popular definition refers mainly to types of embodied individuals, whereas the scientific definition bears specifically on the process of sexual reproduction (while acknowledging an ambiguity of sense). But whether we use Margulis and Sagan's definition or the popular one, neither in fact apply in any straightforward way to plants in the way that they apply to animals. Further, the investigation of the question of plant sex forces us to ask questions about the meaning of 'sex' in general. It is not obvious what it means to say that plants are sexed, and one outcome of the further investigation of this problem is the realization that there are several different meanings of 'sex' in general and that their confusion has profound social and political implications.

The complexity of the question of plant sex is more or less systematically covered over in modern histories of botany, precisely where one may have expected to find it discussed: in the history of the theory of plant sexuality. As we indicated in the Introduction, traditional histories of the theory of plant 'sexuality', written from the standpoint of the present, take the idea of plant sexuality for granted as an empirical fact, search for early indications of its recognition and criticize the thinkers of past for failing to discover the knowledge of the present. A. W. Morton, for example, finds it strange that various otherwise important botanists (like Marcello Malpighi) failed to entertain the idea of plant sex.[2] For Morton, pollen is 'manifestly the male element'[3] in plant reproduction and he struggles to understand why botanists could not accept that straight away.[4] In the most sustained treatment of the topic in English, Lincoln Taiz and Lee Taiz organize their book *Flora Unveiled* (2017) around two 'basic questions': 'why did it take so long to discover sex in plants, and why, after its proposal and experimental confirmation in the late seventeenth century, did the debate continue for another 150 years?'[5] Like Morton, they wonder why it took so long for the past to catch up with the present, surprised that neither Aristotle nor his contemporary Theophrastus, 'despite their considerable resources', advanced the problem of sex in plants: 'The failure of these two luminaries to recognize the more general role of pollination in hermaphroditic flowers has long puzzled historians of botany.'[6]

We are mainly speaking here of the Western botanical tradition. This tradition, in its developed theoretical aspects up until the scientific advances of the eighteenth century, is fundamentally Aristotelian. Traditional histories of the theory of plant sexuality understand the progress towards a properly scientific botany as dependent on its liberation from the philosophical commitments of the early, Aristotelian plant philosophers (especially Nicolaus of Damascus, Albertus Magnus and Cesalpino) and in terms of the rejection of analogies with animals in favour of a literal approach to plant sex and reproduction. Plant science extricates itself from plant philosophy and at last plant sexuality becomes obvious. In effect, according to the traditional histories, 'plant philosophy' was the problem all along.

But approaching the history of the theory of plant sexuality from a positive plant philosophy perspective, in which the meaning of 'sexuality' is not taken for granted, things look rather different. We see that the zoocentric model in the study and understanding of plant life is nowhere more obvious than in discussions of plant sex, which are indeed archetypes of the zoocentric model. We see the extent to which that zoocentrism is the outcome of a history of explicitly analogical reasoning in the study and philosophy of plant life – a history of analogies between animals and plants, with animals supplying the 'source' terms with which plants are compared. But we also see that the question of plant sex (or the question: Do plants have male and female?) cannot be addressed without first asking a more fundamental, philosophical question: what are 'male' and 'female'? We see that plant sex was never 'obvious' for the plant philosophy tradition precisely because it paid attention to this prior, philosophical question. We see how the most important moments in the 'discovery' of plant sexuality are connected to this philosophical question and how the plant philosophers concerned reflected explicitly and critically on the analogy with animals, rather than taking it for granted. In plant science the zoocentric and anthropological terms in which plant sex is discussed now almost always get a free pass, and the conflation of the different senses of 'sex' and 'male' and 'female' sows confusion in discussions of plant reproduction and ecology. This same conflation in popular discourses sows political confusion and leads to the reactionary biologization of social phenomena.

The emergence of the strikingly original works by Marder and Coccia, especially, in the early twenty-first century could give the impression that plant philosophy is a very recent phenomenon, even where it revives ancient ideas (such as Marder's treatment of the idea of the vegetative soul). But in this chapter we will reconnect it with a much longer tradition, drawing out one particular thread – discussions of plant sex – in order to raise the philosophical questions about 'sex' and 'male' and 'female' again in the context of contemporary plant philosophy and science. If, as we have said, discussions of plant sex are archetypes of the zoocentric model, we need to understand the analogical forms that underlie this, and the historical basis of explicitly analogical reasoning is to be found in Aristotle's philosophy. In this chapter, then, we will look at the different kinds of analogy in Aristotle's philosophy generally, at the analogical models employed in Aristotle's zoology specifically and at the idea of the vegetative soul. It is this idea of the vegetative soul, we will argue, that makes analogies between animals and plants possible and which is (as we shall see in the next chapter), contrary to the traditional histories of botany, one of the bases of the developments in the theory of plant sexuality in the eighteenth century. We will identify the specificity of the analogical form in questions concerning male and female in plants as Aristotle poses them, and in the treatment of the same question in the Aristotelian botanical tradition (or plant philosophy) represented by Theophrastus, Nicolaus of Damascus and Albertus Magnus. Perhaps this seems remote from contemporary social and political questions about sex, but it is not. Because it is, perhaps surprisingly, in Aristotelian plant philosophy that we find the reflection on the different meanings of sex that we might do well to reintroduce into contemporary debates.

Forms of analogy and the maintenance of difference

If botany, from its classical sources and well into the eighteenth century, was often based on a series of analogies with animals, this is no doubt in part (if not wholly) due to the overwhelming importance of Aristotelian philosophy in theoretical botany.[7] Aristotle employed analogies throughout his philosophical works and especially frequently in his zoology. In the programmatic early remarks in both *History of Animals* and *Parts of Animals* it is made clear that analogical comparison is a formal methodological principle in Aristotle's zoological investigations. In both works it is stated that within 'genera' (by which he means any group constituted by what its members have in common; this is not the modern taxonomic concept of genus) things differ from each other by degree. Aristotle calls these differences of the 'more and the less', or of 'excess and defect', whereas across genera things differ in kind and any sameness between them is only 'in the way of analogy':

> For all kinds [*genōn*] that differ by degree and by *the more and the less* have been linked under one kind [*heni genei*], while all that are analogous have been separated. I mean, for example, that bird differs from bird by *the more* or by degree (one is long-feathered, another is short-feathered), but fishes differ from bird by analogy (what is feather in one is scale in the other). But to do this for all is not easy, since the similarity in most animals is by analogy.[8]

Analogies pick out specific aspects of animal bodies where there is a commonality of function ('parts' being explicitly correlated with functions[9]):

> By 'analogously' I mean that some have lungs while others have not lungs but something else instead which is to them what lungs are to the former; and some have blood while others have the analogous part that possesses the same capability as blood does for the blooded.[10]

The concept of analogy in modern biology (similarity of function in parts of unrelated genera) is derived with surprisingly little change from this. We can call this a 'formal' analogy, or an analogy of relation (of part[s] to whole) in the different species – in the logical sense – being compared.[11]

Aristotle sometimes points out such analogies where they are relatively obvious, for example: 'The quadruped vivipara instead of arms have forelegs. This is true of all quadrupeds, but such of them as have toes have, practically speaking, organs analogous to hands; at all events, they use these forelimbs for many purposes as hands – except for the elephant.'[12] He also uses analogy, implicitly and explicitly, where the relation between parts of different genera is not so obvious, but where there *must be* analogies as all animals must have the parts that correspond to the functions that define them as animals.[13]

Aristotle also uses what we can call 'substantive' analogy, which works via the identification of shared properties, implying a shared nature. This can be the basis for the definition or identification of a genus.[14] Mary Hesse cites Aristotle's example, in *History of Animals*, of the analogy between (human) bone and fish bone or spine, suggesting that the analogy implies that they share an 'osseous' nature (for which there is no specific word).[15] Finally, Hesse identifies a third, 'metaphysical' kind of analogy. In *De Anima*, for example, Aristotle explains a distinction between two sorts of actuality, the example being the difference between the actuality of knowledge and the actuality of contemplating: 'as waking is analogous to contemplation [both actualities], sleeping is analogous to having knowledge without exercising it [both potentialities].'[16] In the *Metaphysics*, the same philosophical problem – the definition of actuality – is also explained with recourse to analogy, but in a different way. Contrasting actuality with potentiality explicitly, and drawing a series of actual-potential antitheses (e.g. building: capable of building; waking: sleeping; seeing: eyes shut but capable of sight; what is shaped out of matter: the matter), Aristotle says: 'Let actuality be defined by one member of this antithesis, and the potential by the other. But all things are not said in the *same sense* to exist actually, but only by analogy... for some are as movement to potentiality, and the others as substance to some sort of matter.' This is an example of a case in which we must not seek a definition (here, of actuality) 'but be content to grasp the analogy'.[17] The *Physics* supplies us with another such example, concerning 'underlying reality' [*hupokeimenē phusis*].[18] Hesse connects this to the earlier discussion of 'one' in the *Metaphysics*, where Aristotle writes that

> some things are one in number, others in species, others in genus, others by analogy; in number those whose matter is one, in species those whose formula is one, in genus those to which the same figure of predication applies, by analogy those which are related as a third thing is to a fourth... things that are one by analogy are not all one in genus.[19]

In these examples, as Hesse says, analogy is the basis of a kind of 'definition' where definition *per genus et differentiam* is impossible, because what is being defined is not itself either genus or species, or more specifically because there is no higher genus ('actuality', 'one', 'underlying reality', and perhaps all metaphysical terms are not genera). It also applies, as here, when the reduction to univocity is impossible (the term that is 'defined' does not apply to the examples used to explain it in the same way, precisely because the term crosses genera), but where the term is also not absolutely equivocal (having different meanings).[20] This may seem to be far from our topic of vegetal sex, but, as we shall see, 'male' and 'female' in plants are in fact 'defined' in this way in the Aristotelian botanical tradition.

The peculiarity of analogy in all of its forms (formal, substantive and metaphysical) lies in the first instance in its being the method to compare and make demonstrations across genera (in the sense identified earlier: any group constituted by what its members have in common).[21] One important consequence of this, and what is crucial to its legitimate use, is that the terms of an analogy (the things compared) are not rendered *identical* by the analogy; their difference is maintained, despite the postulation of a similarity.

So what kind of analogy is at work in analogies between animals and plants? And what possible basis is there for such analogies in the first place?

The vegetable soul

Aristotle's conception of the hierarchically arranged unity of the natural world assigned vegetable life a place that both connected and separated it from animal and human life. One paragraph from Chapter VIII of the *History of Animals* sets this out clearly:

> Nature proceeds little by little from things lifeless to animal life [*ek tōn apsuchōn eis ta zōia*] in such a way that it is impossible to determine the exact line of demarcation, nor on which side thereof an intermediate form should lie. Thus, next after lifeless things comes the plant [*tōn phutōn*], and of plants one will differ from another as to its amount of apparent vitality; and, in a word, the whole genus of plants, whilst it is devoid of life as compared with an animal, is endowed with life as compared with other corporeal entities. Indeed, as we just remarked, there is observed in plants a continuous scale of ascent towards the animal. So, in the sea, there are certain objects concerning which one would be at a loss to determine whether they be animal or vegetable [*ē phuton*].[22]

This 'continuous scale of ascent', although it may be traced through physical features, is secured philosophically, for Aristotle, through the postulation of the three 'parts' of the soul (*psuchē*) or three kinds of soul: nutritive (or vegetable), perceptive (or sensitive) and rational. Aristotle postulates that soul is a substance as 'the form of a natural body which has life in potentiality'. By 'life', he says, he means 'that which has through itself nourishment, growth, and decay'.[23] A natural body that has life in potentiality is one which is 'organic', which means both having organs and, as is later clarified, that the body is an organ of the soul.[24] In the next chapter, living things are defined as those that have at least one of the following: 'reason, perception, motion and rest with respect to place, and further the motion in relation to nourishment, decay and growth'. As plants have the latter 'motion', 'even plants, all of them, seem to be alive'.[25] The part of the soul corresponding to this motion is the 'nutritive capacity' (*threptikon dunatai*). This also came to be known, via the Latin, as the 'vegetative' part of the soul. This nutritive capacity is the principle through which all living things are alive. All animals, in addition, have perception or sensation (which implies also desire, appetite, spirit, pleasure and pain);[26] some (i.e. human) animals, as well as the nutritive and perceptual faculty, also have the faculty of understanding and reason (*to dianoētikon te kai noûs*).[27] Thus the 'nutritive soul' (*threptikē psuchē*) is 'the first and most common capacity of the soul in virtue of which living belongs to all living things, a capacity whose functions are generating and making use of nutrition'.[28]

There is a kind of substantive analogy at work here, to the extent that the comparison of nutrition and generation in plants and animals implies a shared nutritive nature giving rise to the definition of the vegetative soul. And this shared organic, natural being and the shared principle of nutrition, growth and generation are then itself the basis for any formal analogy between plants and animals. This is evident in Aristotle's first such analogy, which is also the first mention of plants in *De Anima*: 'And even the parts of plants are organs, although altogether simple ones. For example, the leaf is a shelter of the outer covering, and the outer covering of the fruit; and the roots are analogous to the mouth, since both draw in nourishment.'[29] The shared nutritive capacity (or shared nutritive soul) is the basis of the presumption that there must be something in plants analogous to the parts of animals that serve the nutritive function.

The basic formal analogy between mouth (and head) and roots is made several times, but the inexactness of the analogy is shown by its variants and by the fact that anything more than the vaguest comparison of the nutritive parts is quickly strained.[30] Indeed, because plants have fewer functions and correspondingly fewer organs than animals (fewer heterogeneous parts), analogies between them are inevitably very limited and Aristotle concludes that 'the configuration of plants is a matter then for separate consideration.'[31] Even here, then, in what is arguably the beginning of the zoocentric tradition in discussions of plants, the inadequacy of animal models and the specificity of plant forms is already implicitly admitted.

What are 'male' and 'female'?

In principle, the shared vegetative soul in animals and plants, associated with the functions of nutrition, growth and generation, means that, for Aristotle, there must also be analogies between plant and animal *generation*. In *History of Animals* Aristotle notes that a common property of plants and animals is that in both groups some are generated from seed/parent and some are spontaneously generated.[32] In *Generation of Animals* an analogy is drawn between animals' production of young and plants' production of seed and fruit, as if the seed were the offspring, but more often the analogy is between plant seed and semen, and Aristotle admits that 'There is considerable difficulty in understanding how the plant is formed out of the seed or any animal out of semen [*tou spermatos tōn phutōn ē tōn zōiōn hotioun*]'.[33] He also compares animals' eggs to the seeds of plants, and in so doing broaches the difficult issue of the applicability of the male/female division to plants. However, in this instance the form of the analogy is significantly different from almost all of Aristotle's other zoological analogies. As we have noted, Aristotle's zoological analogies (between animals) are primarily formal analogies between functionally similar parts. Such analogies between functionally similar parts of animals and plants are from the beginning more difficult in the case of male and female because of the lack of any obvious similarity between any parts, and the complete absence, as far as Aristotle is concerned, of any specifically male and female *parts* in plants. This means that the analogy between 'male' and 'female' in plants and animals will of necessity take a completely different form.

Some modern historians of botany find it hard to understand why Aristotle did not recognize the 'sexuality' of plants.[34] But when we understand what, for Aristotle, 'male' and 'female' are and what roles they play in reproduction, we can see that the idea of male and female plants posed a particular problem for him. In *Generation of Animals* Aristotle's aim is to explain both the specific contributions of male and female to generation and the generation of male and female animals. In this context the male and the female are defined, primarily, in terms of difference in capacity, role or function:

Male is what we call an animal that generates into another, female that which generates into itself.... Male and female differ by definition [*logon*] in having different capabilities [*dunasthai*], and by appearance in certain parts. They differ by definition in that the male is that which can generate into another... while the female is that which generates into itself and out of which the generated offspring is produced while present in the generator. Since they are distinguished by capability [*dunamei*] and by particular function [*ergōi*], and since there is need of organs for every performance of function, and the organs for the capabilities are the parts of the body, there must be parts both for producing young and for coupling, and these parts must be different from each other. For although male and female are predicated of the animal as a whole, nevertheless it is not male or female in respect of all of itself but in respect of a particular capability and a particular part.[35]

As is well known, Aristotle claims that the female contributes the matter, while the male imparts the motion and the soul in generation. And this is another way, indeed, of *defining* the male and the female: 'the power of making the sensitive soul' is the essence of what being male is – it is what it is to be a male.[36] Aristotle makes this latter claim in the context of the problem of why animals need males to reproduce, given that females possess the same soul as males and provide the material for the embryo:

> The reason is that the animal differs from the plant by having sense-perception [*aisthēsei*]; if the sensitive soul [*aisthētikēs psuchēs*] is not present, either actually or potentially, and either with or without qualification, it is impossible for face, hand, flesh, or any other part to exist; it will be no better than a corpse or a part of the corpse. Thus if it is the male that has the power of making the sensitive soul, then where male and female are separate it is impossible for the female to generate an animal from itself alone, for the process in question was what being male is.[37]

If 'being a male' just is the power of making the sensitive or perceptive soul (in English translations 'sensation' and 'perception' both translate *aisthēsis*) there cannot by definition be any male plants, because no plants have such a power. Although Aristotle must admit that plants have parts of some kind (as we shall see, what their parts are is a very difficult question for him and his contemporary Theophrastus), there can be for him no 'sensitive' parts, which we must assume to include the sexual parts.[38]

Together, Aristotle's definition of an animal and his definition of a male seem to contradict his claim that some animals, however, have no male and female.[39] How can this be the case? As Aristotle himself asks, what in these cases corresponds to the female material principle that *Generation of Animals* takes to be located in the residue secreted by female animals, and 'what is the moving principle which corresponds to the male?'[40] There are in fact two problems here. First, how are such animals still animals, if they do not have the male that imparts the sensitive soul? Some remarks suggest that such 'animals' might represent the point of indifference between animal and plant, for example the testacea: 'because their being is almost like plants there is no male and female any more than in plants.'[41] Some testacea, according to Aristotle, are formed spontaneously while 'some emit a sort of generative substance from themselves', but this is not 'real semen [*sperma*]'. To understand the generation of such creatures we need to 'grasp the different methods of generation in plants; some of these are produced from seed, some from slips, planted out, some by budding off alongside'; testacea 'participate in the resemblance to plants' in this manner.[42]

However, as all living things are defined as those that have 'motion [*kinēsis*] in relation to nourishment, decay and growth' and the male is said to be that which imparts movement (more generally; not specifically the sensitive soul) to the female material principle,[43] there can be no living thing at all – animal or plant – that does not have the male principle. It seems, then, that what Aristotle means when he says that some animals have no male and female is that in some animals there is no *separation* between male and female as 'male and female are distinct in the most perfect [animals], and... these [male and female] are first principles [*dunameis archas*] of all living things whether animals or plants, only in some of them these principles are separated and in others not'.[44] More specifically, all plants and some animals contain the male principle within themselves mixed with the female. Aristotle sometimes seems to suggest that in such cases (e.g. with bees and certain fishes) it may appear that the female generates without the male, but in fact the 'female' contains the male principle within herself 'as plants do' and this is sufficient, without the contribution of a *separate* male, for 'fetation' to occur with the relation between nourishment and heat.[45] In sum:

> In all animals which move about, male and female are separate; one animal is male and another female, though they are identical in species, just as men and women are both human beings, and stallion and mare are both horses. In plants, however, these faculties [*dunameis*] are mingled together; the female is not separate from the male; and that is why they generate out of themselves, and produce not semen [*gonēn*] but a fetation [*kuēma*] – what we call their 'seeds'.[46]

Whatever these male and female 'faculties' or 'principles' are, they are obviously not identifiable with male and female *individuals* or male and female *parts*, not least because they are said to be the cause of male and female individuals and male and female parts.[47] This suggests that there are several senses of 'male' and 'female' in Aristotle's remarks on living beings, and that 'male' and 'female' are defined according to what Hesse calls a metaphysical analogy, as they are neither univocal terms ('male' and 'female' are not always meant in the same sense) nor absolutely equivocal. 'Male' and 'female' are (i) *metaphysical principles* of the generation of life which may be found mixed together in one plant or animal or found in separate individuals, (ii) separate kinds of *individuals* in some kinds of animals and (iii) separate kinds of *parts* of individuals in some kinds of animals. When Aristotle says that there is no male or female in plants he means no male or female in senses (ii) and (iii). Aristotle concedes that individual plants are sometimes *called* 'male' and 'female' in sense (ii) – a particular kind of plant may be said to divided into 'male' and 'female' individuals – but only by a kind of analogy. In *Generation of Animals*, for example, he says that, although there is no male and female in plants and some animals (e.g. testacea),

> they have come to be called male and female in virtue of resemblance and analogy, for they [animals like testacea] have a small differentiation of this sort. Among the plants too there are in the same kind of tree some that are fruit-bearing and some that do not bear themselves but contribute to the concocting of others' fruit, as occurs with the fig and caprifig.[48]

But Aristotle does not ever suggest any analogy of male and female *parts* (sense iii) or of functions of parts – that is, no analogy of the form that is otherwise common and indeed methodologically central in his zoology, and (as we shall see in later chapters) is the form that dominates the discussion and understanding of male and female in plants from the seventeenth century to the present day.

However, Aristotle's hylomorphism (the metaphysical doctrine that all beings are composed through two principles, form and matter) and the speculative metaphysics of generation based on it require him to think male and female principles, as principles of life, in sense (i), even in plants. Every plant of every kind, just because it is a living thing, must be understood to contain within in it – ostensibly in reality, not analogically – male and female principles. But what does this mean – 'male' and 'female' *principles*? Discussion of the male and female principles in the literature on Aristotle tends to approach the question with the idea of (separate) male and female animals (sexed individuals) in mind, most often human animals.[49] But the problem is really to identify and explain what male and female principles are across all of the animal kingdom, including where they are not separate, *and in plants*. Indeed, the need to answer the question in relation to living beings where (according to Aristotle) 'male' and 'female' are not separate and there are no 'male' and 'female' parts means that the question 'do plants have male and female?' becomes, of necessity, a more fundamental question about the nature of the metaphysical principles of male and female themselves.

As we shall see below, this becomes more explicit in the later Aristotelian texts that deal with the question of male and female in plants, especially those of Nicolaus of Damascus and Albertus Magnus. But let us just note, for now, the result of the plant-philosophical orientation of our discussion of Aristotle. The analogies between animals and plants in Aristotle's work have been shown to be mostly of the formal type (analogies of part and function across genera), and these have been shown to rest on a kind of substantive analogy of capacities that gives rise to the definition of the vegetative soul. As Aristotle understands the functions of nutrition, growth and generation – precisely, the functions of any living organism – in terms of organs or parts, his analogies try to identify the relevant organs and parts of plants, even as he recognizes that plants do not, in fact, have 'parts' in the way that animals do. This general point is nowhere more clearly articulated than in the attempt to find analogies in the generation of animals and of plants. Indeed, this of necessity leads Aristotle away from analogies of parts to the more fundamental, metaphysical question: what are 'male' and 'female' such that these principles can be said to be part of vegetable life? As the answer requires us to distinguish between different senses of 'male' and 'female' these terms are shown to be defined according to what Hesse calls 'metaphysical analogy'. Like all metaphysical terms they are neither univocal nor absolutely equivocal, and the philosophical challenge is to maintain their different senses while understanding the nature of their generality. Thus, the discussion of what will later be called plant 'sex' turns out to be of greater philosophical interest than its restriction to botany might have us believe.

What we call 'male' and 'female'

In the surviving texts and fragments that comprise the Aristotelian tradition in botanical philosophy the question of male and female in plants is considerably developed – indeed, it becomes one of the central questions. It is interesting, however, that Aristotle's contemporary Theophrastus, whose two books on plants (*Enquiry into Plants* and *Causes of Plants*) echo the form and procedure of Aristotle's zoology in many ways, studiously avoids all question of the metaphysical principles of male and female. *Enquiry into Plants* begins (like *History of Animals* and *Parts of Animals*) with methodological considerations, and the extent of the validity of plant-animal analogies is at the heart of these: 'In considering the distinctive characters of plants and their nature generally one must take into account their parts, their qualities, the ways in which their life originates, and the course which it follows in each case: (conduct and activities we do not find in them, as we do in animals).'[50] But we are immediately faced with the problem that it not obvious what should count as a 'part' of a plant. As something belonging to a plant's characteristic nature it seems that a 'part' should be something that, once it has appeared, is permanent – like the parts of animals. But many of the parts of plants, and especially those concerned with reproduction, are ephemeral. If we count these the number of parts will be indeterminate and constantly changing; if we do not count them essential features of the plants will be left out.[51]

Theophrastus's answer to this stresses the need to be cautious with any plant-animal analogy while also suggesting that it will often be most fruitful to proceed precisely by such comparisons. In some cases, where there exist no special names for particular parts of plants, resemblance (of part and of function) dictates that the names of corresponding parts of animals are borrowed – for example 'fibre' and 'vein'.[52] This and other analogies with animals are justified because, in general, animals are simply better known, and it is by the better known that we must pursue the unknown. Analogies concerning generation are identified as particularly challenging:

> But perhaps we should not expect to find in plants a complete correspondence with animals in regard to those things which concern reproduction any more than in other respects; and so we should reckon as 'parts' even those things to which the plant gives birth [*ta gennōmena*], for instance their fruits, although we do not so reckon the unborn young of animals.[53]

Does Theophrastus pursue an analogy between male and female in animals and plants? Certainly, he speaks relatively often of male and female plants (principally trees, his model for all plants), but it is striking that this is not part of any functional analogy. Rather, 'male' and 'female' are most often for Theophrastus folk-designations for different types of plants of what we would call the same species. As such, he often notes that what counts as 'male' and what counts as 'female' is differently determined in different contexts and that these multiple determinations are sometimes incompatible. Some say, for example, that in plants of the same kind the male flowers while the female does not.[54] Some say males do not bear fruit but others say that only males do.[55] But however they are identified, the difference between 'male' and 'female' is, according to Theophrastus, 'of the same character as that which distinguishes the cultivated from the wild tree, while other differences distinguish different forms [eidos] of the same kind [homogenōn]',[56] and the difference between fruitless and fruit-bearing, flowering and flowerless (which can also be markers of 'male' and 'female') is said to be due to the position of the plant and the climate of the district.[57]

In Book II of *Enquiry into Plants*, when Theophrastus considers propagation (spontaneous growth, growth from seed, root, a piece torn off, from a branch or a twig, from the trunk and so on), there is no attempt to make any functional analogy between male and female animals and male and female plants.[58] The only exception – and the most frequently cited example of male and female trees in the ancient literature – is date palms: 'With dates it is helpful to bring the male to the female; for it is the male which causes the fruit to persist and ripen.' He describes the process of shaking the 'dust' of the male flower over the female fruit, allowing the male to 'render aid' to the female – 'for the fruit bearing tree is called "female"'.[59] But even here Theophrastus mentions no possible analogy with animal generation, or with the nature of the distinction between male and female animals.

In Book III, the differences between so-called male and female trees in general are said to concern the qualities of the timber. According to the woodmen any wood of a male tree cut with an axe 'gives shorter lengths, is more twisted, harder to work, and darker in colour', while the female gives better lengths and contains the heart wood that is less resinous and soaked with pitch so smoother and of straighter grain.[60] In various trees the wood of the female is said to be whiter, softer and easier to work; the male thicker, harder, unbending and inferior in appearance.[61] Thus if there is an analogy between male and female in the animal and the plant worlds, it is not one based on any attempt to find analogous functions in generation. Rather, it is an analogy based on the identification of common properties assumed to be characteristic of male and female animals, and perhaps male and female humans above all.[62]

In *Causes of Plants*, devoted to generation and propagation, we again find no functional analogy between male and female animals and plants (even though Theophrastus speaks more than once of 'pregnant' trees).[63] The difference between male and female is once again understood to be qualitative, and Theophrastus is careful to play down the analogy with animals. Perhaps the fact that wild trees produce more seed is explicable in terms of their

> being as it were male [*hōsper arrena onta*] and in their nature closer in texture and drier, for cultivating and good feeding have an effeminating effect [*diathēlunousi gar hai katergasiai kai hai trophai*]. This remark however, is to be taken as resting on a certain resemblance, and the parallel is pretty remote [*touto men ouv hōs kath' homoiotēta tina legesthō, porrōter ō keimenon*].[64]

If female trees are more fruitful than males they are nevertheless not so hot, 'as one can assume from their similarity to the female in animals, even if here the word has a different sense'. The relation between the male and the female date palm is said to be 'something similar' to the relationship between fish, 'when the male sprinkles his milt on the eggs as they are laid. But resemblances can be found even in things widely separate'.[65] In sum, then, Theophrastus treats male and female as popular names for types of plants (principally trees), and his descriptions suggest that the popular designations are based on an analogy between some of the imagined properties of male and female animals, but not on their roles in reproduction.[66] The metaphysical principles of male and female play no role in Theophrastus's discussion.

However, the subsequent, and more overtly philosophical tradition in botany foregrounds the metaphysical aspect. The first-century BCE *De Plantis* of Nicolaus Damascenus is thought to be a re-elaboration of Aristotle's lost work of the same title and was until relatively recently misattributed to Aristotle.[67] Only fragments of translations of the lost Syriac translation of Nicolaus's *De Plantis* survive, but the various translators who also commented on the texts elaborate on it for us. These texts seem to say contradictory things about the possibility of male and female in plants, which they acknowledge is an important question (along with the questions of whether plants have souls and/or sensation, whether they breathe and whether they sleep). All of the different translations frame the question as a response to a claim attributed to Empedocles (translated here from Bar Hebraeus) that plants 'are males and females at the same time, because in each of them these two natures are mingled and united'.[68] Isḥāq ibn Ḥunayn's text says (in translation, at least), 'I wish I knew whether plants... have males and females, or something in which male and female are combined',[69] suggesting two options. Some note the objection that the latter seems to presuppose the former, because that which is mingled must first have been separate.[70] Most then suggest that the answer to this question lies somehow in the comparison (which Ibn At-Tayyib attributes to Aristotle[71]) of birds' eggs and seeds. Bar Hebraeus's commentary, in his *Butyrum Sapientiae* (*The Cream of Wisdom*), is most interesting on this point:

If a reader calls a body male that by nature has a capacity for moving a matter to receive a form similar to its own, and female a body that has a passive capacity for receiving a form in the above mentioned way, then plants have male and female, and it is possible that a single plant is male and female *at the same time*: female because the said matter is generated in it, and male because the said moving capacity originates in it.

But if it is not so, but the reader calls a body male from which a matter exudes by means of organs intended for that, which acts upon a matter present in the recipient, and he *calls* the recipient female, then there is neither male nor female in plants.

Let us agree, then to call female the capacity that secretes a residue from the plant that enters into the constitution of the like generated out of it, and male the forming capacity that shapes the residue.[72]

In animals, he continues, these two capacities are united in the foetus after the intercourse of two separate individuals. Similarly, in every egg 'there is a generative capacity and a capacity for receiving generation; and therefore a chick is generated in each of them. The process is similar to that of seeds in plants'.[73] Bar Hebraeus seems, in this way, to make a distinction between, on the one hand, male and female as separate individuals or kinds, one (male) exuding a matter into the other, the recipient (female); and, on the other hand, male and female as capacities (the capacity to give and receive form respectively) which may both be inherent in one individual – a distinction that would resolve the apparent contradictions in *De Plantis*. Plants do not have male and female in the first sense, but they do in the second.

Ḥunayn's text similarly denies that plants have separate male and female in the first sense, as 'simple components' that become mingled, while affirming that they do have male and female in the second sense:

> But as regards the mingling of males and female in plants, we must understand it differently, for the seed of a plant resembles a fetation, and that is a mingling of male and female. And just as in an egg there exists the power to generate a chicken, and *along with it* the material to feed it up to the time of its completion and its hatching, while *on the other hand* the female lies [sic] one egg at a time, so *it is with the seed* of a plant.[74]

These different senses seem to correspond to the distinction in Aristotle's works between male and female as separate kinds of individuals and male and female as metaphysical principles. In animals the metaphysical principles reside in the different individuals, to become mixed in intercourse and its products (the foetus and the egg). In plants the metaphysical principles are already found mixed in the plant, or more precisely in the seed, which in this respect resembles an egg.[75] In Bar Hebraeus's commentary these metaphysical principles of male and female are based on the more fundamental distinction between activity and passivity, making explicit what we have argued is implicit in Aristotle's texts.

This is also the case with Albertus Magnus's thirteenth-century discussion of the problem of the sex of plants in his *De Vegetabilibus*. Albertus explicitly claims to resolve the contradictions on this question that arise from a survey of the ancients, principally the contradiction between Aristotle (who, according to Albertus, says there is no male and female in plants) and Theophrastus and others (who identify various male and female plants) or say that male and female are mixed in plants (a position that logically requires them to have been separate before they are mixed). Albertus effectively identifies the cause of these contradictions in the possibility of different ways of speaking about male and female in relation to plants and animals.

If, he writes, we define 'male' as that which from its own seed generates in another, and 'female' as that which takes seed from another and generates in itself, there can be no doubt but that 'sex is not found in plants'.[76] Thus Albertus affirms Aristotle's definitions of male and female in animals and the consequence of them for the possibility of sex in plants. But is it possible, he continues, that some shared sexual properties or characteristics (*aliquas proprietates sexum participantium*) can be found in plants? Albertus then lists some of the qualities that Theophrastus had used to describe male and female plants (though he does not name Theophrastus), with the addition of an Aristotelian explanation for these different qualities:

> So, the masculine, because of its forming and stamping [role] is harder and drier and as a consequence rougher to touch. The feminine however, because it is formed and it is stamped, has properties the opposite of these, so softness, moistness, smoothness, since those properties do well to bear the act of forming. And we find in plants that are called masculine that every one that is reproduced from these is stronger and rougher. And every one that is generated from those which are called female is softer and smoother.

His conclusion – it is thus clear that 'the sexes may not be distinct in plants'[77] – may be contested by the reader confused by the transfer of the definitions of male and female (forming and stamping/formed and stamped) into plants, but it is only the qualities that are *associated with* the roles of male and female that are transferred to those plants called male and female, not the roles themselves.[78]

A few chapters later, Albertus returns to those ancients who say that there are male plants and female plants (e.g. olives and peonies). These authorities assert that the generation of every living thing involves a power proper to that thing (*agens proprium*), and that the active power cannot be the same as the passive 'power' because that would involve the identity of opposites (principally, action would be the same as passion). According to this view (which Albertus associates with Plato), 'In plants therefore active generation and passive generation in a single species will be separated by substance'.[79] Albertus counters this view with Aristotle's claim that male and female are inseparable accidents peculiar to animals, and not plants,[80] and his claim, in his book on animals (presumably *Generation of Animals*), that the distinction of the male sex from the female sex is only to be found where there is a sensitive soul. If it was only the vegetative soul that was responsible for generation (in animals, presumably) no distinction of sex would be necessary.[81] The power responsible for the sensitive soul cannot be in the passive 'seed' of the female, as the phenomenon of wind eggs shows, so in the more perfect generation of animals (more perfect than plants, that is), something in addition to the female seed is required to contribute the sensitive soul, whereas (it is implied) this is not necessary in the plant. Thus, Albertus says, it is Aristotle's argument that plants do not combine (*conveniat*) male and female in generation, and if the contrary position is said to be attributed to him, it can only be that he means it metaphorically.[82]

This metaphorical sense is further explained in the next paragraph. There is a true sex or true power or virtue (*virtus*) of sex and an imitation of this virtue or power. The true sex (one sex 'forming and creating' the sensitive soul in the animal seed, the other 'worked and formed' in the organ of its soul) is 'by no means in plants'. The action of the sun on the seed of a plant is analogous (*similtudo*) to the action of the male power of sex on the passive female seed, but this is the only sense in which there can be said to be sex in plants, 'not a true version, but a distant imitation'. This is reiterated at the end of the chapter: 'It is therefore the conclusion of this question, according to the Aristotelian tradition, that sex in no way combines [*convenit*] in plants, in the same way as it combines in animals' but there is a 'remote analogy' (*similitudinem remotam*), and this distinction between true sex and its remote analogy resolves the contradiction in the disagreements of the ancients on this question. In whatever way the ancients conceived of the distinction of sex in animals, it is only by analogy that this distinction can be said to apply to plants.[83] Albertus thus goes further than Nicolaus in denying that plants have male and female, because he denies not only that they are male and female *individuals* or substances (separate or mixed) in any real sense but also that the metaphysical principles (or 'virtues') of male and female can be said to belong to plants even analogically.

Once again, we see that the question of male and female in plants cannot be addressed by these authors without tackling the more fundamental question about the nature of the metaphysical principles of 'male' and 'female' themselves. Or, to put this in another way, the very difficult question of what we mean by 'male' and 'female' across genera, across different forms of life, is forced to philosophical attention by the study (albeit it is here exclusively theoretical) of plants. It is not that the meaning of 'male' and 'female' is clear, and that the problem is only to discover whether there are male and female plants. It is not, then, an empirical problem. Rather, the life of plants complicates the question of what is meant by 'male' and 'female', giving rise to a philosophical problem. This, in a nutshell, is the central thesis of this book.

When we focus on this philosophical question, it is hard not to conclude that the positing of male and female *principles* in all living things is itself the result of an implicit analogy with the generation of animals divided into male and female *individuals*. That is, presuming that there must be active (animating) and passive (receiving) principles in generation, these principles are called 'male' and 'female', borrowing the terms of a distinction between kinds of *individuals* in the animal kingdom. This provides the metaphysical definition of 'male' and 'female', terms which, in their generality, are applied to all forms of life, but only insofar as their various, different senses are respected when speaking of specific instances. With a complicated self-justifying circularity, the *result* of an implicit analogical (or perhaps metaphorical) extension of the animal division between male and female individuals to the presumed active/passive metaphysical principles of generation gives us the idea of male and female principles, which are then evoked as the metaphysical basis for the analogy between animal and plant sex.

What is the justification for the characterization of the different principles or causes (of motion and matter, respectively) as 'male' and 'female'? What is the justification for the characterization of activity and passivity as male and female respectively? There is no explicit justification of this; these characterizations are, rather, the foundational presumptions of the metaphysics of 'sex'.

3 THE JOINT VENTURE: PHILOSOPHY AND BOTANY

At the end of Chapter 2 we saw that the discussion of the specific question of male and female in plants in early plant philosophy could not be addressed without tackling more fundamental questions about the nature of 'male' and 'female' in general, and that in plant philosophy from Aristotle to Albertus these questions are fundamentally metaphysical. In other words, it is through thinking about plants, specifically, that the difficulty of the basic philosophical question 'What are "male" and "female"?' comes to the fore.

Traditional histories of botany, however, see things completely differently. Writing from the standpoint of the alleged 'obviousness' of plant sex in modern science, there is for these histories no problem with the idea of male and female in plants, and certainly no philosophical issue. For them, the problem in the history of the theory of plant sexuality is plant philosophy itself, or philosophy more generally. Philosophy prevents science from seeing what is obvious; the decisive advances in the theory of plant sexuality are made when botany becomes properly scientific in liberating itself from its Aristotelian baggage. As R. J. Harvey-Gibson wrote, Aristotelianism had been extraordinarily influential but by the end of the seventeenth century 'the long night was passing and the dawn of a new day was at hand'.[1]

There can be no doubt but that the development of the theory of the 'sexuality' of plants, and the understanding of the fact and mechanisms of fertilization in plant sexual reproduction, is tied up with the conceptual, methodological and technological innovations that are generally thought to usher in the era of a properly scientific botany. These innovations include attempts at a systematic classification of the plant world, experimental method and – via the new technology of microscopy – the establishment of the science of plant anatomy and plant physiology. Concerning plant classification, Theophrastus had identified what he took to be the basic groups of plants: trees, shrubs, bushes (or under-shrubs) and herbs. These classical distinctions continued to be followed well into the eighteenth century and the words (if not quite the definitions) are still in use today.[2] Albertus added fungi to Theophrastus's groups and some find more sophisticated forms of distinction in *De vegetabilibus*,[3] but Andrea Cesalpino's *De plantis* (1583) is taken to be a significant scientific development, especially because of his classification of plants according to the organs of 'fructification'.[4] Although Cesalpino did not think of these as 'sexual' organs, others after him would. In the decades that followed, Marcello Malpighi (1628–94) and Nehemiah Grew (1641–1712) were amongst the first to use the new technology of microscopy, enabling detailed examination of the inner structures of plants as well as their outer forms. The observations described and illustrated by Malpighi and Grew are now seen as the beginning of plant anatomy and plant physiology. Grew, in particular, examined pollen and the flowering parts of plants and one sentence from Grew's *The Anatomy of Plants* (1682), which is generally understood to be an identification of the plant stamen with the animal penis, is often taken to mark the beginnings of the recognition of plant sex in the literal, non-analogical or non-metaphorical sense. Finally, Rudolf Jakob Camerarius (1665–1721) described, in his famous *De sexu plantarum epistola* (1694) (*Letter on the Sex of Plants*) the experiments that showed that both stamens and anthers are necessary for the formation of fruit and on this basis argued for the necessity of both male and female for plants, as for animals. Whether the 'discovery' of the sex of plants is attributed to Grew (an Englishman) or Camerarius (who was German) seems ultimately be a question of national preference.[5] But all histories of botany are agreed in seeing the late seventeenth century as the turning point in the theory of the sexuality of plants. Thereafter the theory was popularized by Sébastien de Vaillant and, influenced by him, by Linnaeus, who made the sexual parts of plants the basis for his famous sexual system of classification.

These new scientific developments – the focus on the parts of fructification in plants as the basis for a natural classification; close and even microscopic observation of these parts, the identification of these parts as literally (not analogically) sexed and the experiments that proved the need for both 'male' and 'female' – are contrasted in the traditional histories of botany with those aspects of the Aristotelian plant philosophy tradition that are seen as an obstacle to properly scientific knowledge: philosophical commitments and speculation, the alleged denial of the sex of plants and animal-plant analogies. Although it is admitted that traces of the Aristotelian inheritance survive in all of these thinkers, the traditional histories mark them as placing clear water between scientific botany and plant philosophy or philosophy *tout court*. The latter is then thought to have no more legitimate role to play in the former.[6]

In this chapter we will contest the terms of these traditional histories by showing the extent to which some of the major contributors to this crucial period in the history of botany – especially Cesalpino and Grew – relied on Aristotelian (or quasi-Aristotelian) philosophy in the development of their scientific work: that is, the extent to which philosophy enabled, rather than deformed, early scientific botany. We will show the role of the idea of the vegetative soul in Cesalpino's botanical works, including his identification of the fructifying parts as the proper basis for classification, and his commitment to the metaphysical idea of male and female. Re-uniting Grew's philosophy with his botany, we will contest the idea that his comments on the sex of plants are the result of a rejection of philosophical deduction in favour of empirical observation, suggesting instead that they follow quite consistently from his attempt, via a philosophy of life, to apply a broadly Aristotelian model of animal generation to plants. Of course, 'philosophy' and 'science' were not distinguished as academic disciplines until much after Grew. If, retrospectively, we distinguish his philosophical and his botanical works on the basis of their primary content, this should not blind us to the fact that they were for him essentially part of the same intellectual whole. Grew, indeed, provides an exceptionally good example of the extent to which philosophy and botany were partners in a joint venture, an example that helps us to rethink the expulsion of philosophy from scientific botany today.

Andrea Cesalpino: The principle of life and the principle of foetification

Cesalpino's *De Plantis* (1583) was a revival of Aristotelian – that is, explicitly philosophical – botany.[7] His classification of plants according to the organs of fructification is ultimately based on Aristotle's theory of the soul, the relevant aspects of which are set out on the first page of *De plantis*:

> As the nature of plants possesses only that kind of soul by which they are nourished, grow, and produce their like, and they are therefore without sensation and motion in which the nature of animals consists, plants have accordingly need of a much smaller apparatus of organs than animals... But since the function of the nutritive soul consists in producing something like itself, and this like has its origin in the food for the maintaining the life of the individual, or in the seed for continuing the species, perfect plants have at most two parts, which are however of the highest necessity; one part called the root by which they procure food; the other by which they bear the fruit, a kind of foetus for the continuation of the species; and this part is named the stem [*caulis*] in smaller plants, the trunk [*caudex*] in trees.[8]

Cesalpino believed that the classification of plants should be based on the number, place and shape of the functionally most important parts, and that to identify these parts required an understanding of the 'essence' of plant life: 'We look for those similarities and differences which make up the essential nature of plants [*Plantarum substantia*], not for those which are only accidental to them.'[9] Although, of course, the vegetative soul is not specific to plant life (all living beings are possessed of such a soul), it is specific to the plant that has only this kind of soul, and thus the essence of the plant is to be found in the functions of nutrition and generation.[10] As he explained in a letter to his patron Alphonso Tornabuoni in 1563, 'I have therefore classified plant species into genera according to the various methods of producing seeds or something corresponding to seeds in the capacity for reproduction, and according to their similarities, since these most appropriately show the quality of soul through which they acquire the totality of their essential nature'. In this same letter Cesalpino explicitly attributes his success in achieving progress in classification to the fact that he combines 'botanical expertise with philosophical studies, without which no advance can be made'.[11]

Cesalpino was, in general, somewhat ambivalent about analogies between animals and plants. In the first chapter of Book I of *De plantis* he notes a 'slight similarity' between them – for example, as veins draw food from the stomach of the animal so the roots of a plant draw food from the earth – and he invokes the Aristotelian idea of the plant as an upside-down animal to claim that the root is (like the head of the animal) the superior part.[12] However, in most respects, he says, the differences between plants and animals are great. Even so, various questions are pursued via analogy, for example: 'Whether any one part in plants can be assigned as the seat of the soul, such as the heart in animals, is a matter for consideration.' Cesalpino argues that the same soul must be present in both the root and the stem (there are not two separate souls for the nutritive and generative capacities), though somehow distributed between them, such that each part functions only according to its peculiar capacity. As it is rational, he says, to presume that in plants, as in animals, the 'principle of life' (*vitalia principia*) is hidden in the inner parts, it follows that the principle of life (or the soul) in the plant is in the pith (*medulla*) hidden within the stem: 'That this was the opinion of the ancients we may gather from the name, for they called this part in plants the heart [*cor*], or brain [*Cerebrum*] or matrix [*matricem*], because from this part in some degree the principle of foetification (the formation of the seed) is derived.'[13]

For Cesalpino the final end of the plant, as of the animal, was the generation of its like; the most perfect form of generation for the plant was generation from seed. Like animal seed, plant seed is nutritive substance concocted by heat from the 'heart', in the pith. According to Cesalpino, this means that it is not necessary 'to separate a special fertilising substance from the rest of the matter in plants, as it is separated in animals which are thus distinguished as male and female.'[14] Cesalpino recognizes that some sterile plants are popularly called male, and the fruiting varieties female, but, like Theophrastus, he saw this as a superficial analogy of qualities.[15] In his *Peripatetic Questions* (1571), discussing the generation of animals and the contributions of the male and the female, Cesalpino claims (contra Aristotle) that the female contributes both form and matter to her offspring. It looks like he contradicts himself when he claims that this is seen in plants, where there is no separation between male and female,[16] but Cesalpino's (Aristotelian) point is that the metaphysical 'male' and 'female' *principles* of life can be understood to play their role in the generation of plants, without there being separate male and female *individuals*. The plant is only endowed with the vegetative soul and yet it does contribute its form to its offspring, so it is only where the male must, in addition, contribute the sensitive soul that male and female need be separate.

This Aristotelian understanding of male and female as, primarily, metaphysical principles of life is also at the heart of Adam Zalužanský's chapter 'De sexu plantarum' in his *Methodus* of 1592.[17] Zalužanský distinguishes the male from the female in terms of the activity of the former and the passivity of the latter. He uses Aristotle's distinctions to define the male as the sex that 'generates in another plant' and the female as the plant 'that generates in itself'. In some plants, he writes, the principles of male and female are commingled; in others the male and female are separate and the females do not generate without the males. The male 'bristling with leaves erected, impregnates them [the females] by his exhalation and by the mere sight of him and also by his pollen'.[18] As Julius Sachs points out, Zalužanský explains that the commingling of male and female in plants compensates for the fact that (most) plants cannot move, allowing the male to play its active role even so.[19] In his example of the plants with separate sexes, the famous date palm, the plants do actually move, as the females 'bend and bow towards him [the male] with more attractive foliage'.[20] But here, as well, it is primarily in terms of the male and female *principles*, presumed to be a universal feature of life, that Zalužanský conceives of 'sex'. Zalužanský differs from Aristotle only in his willingness to see these principles as being able to be embodied in different individual plants in an attempt, perhaps, to give more substance to the distinctions such as those of the date palm. While Cesalpino, on the contrary, denies that plants have *separate* male and female he too holds fast to the idea of male and female *principles* mingled in the seed or fetation.[21] Cesalpino's and Zalužanský's botany progresses through, not despite their philosophical commitments.

Nehemiah Grew and the plant penis

In the traditional histories of botany it is recognized that Cesalpino remained attached to Aristotelian philosophy while simultaneously playing a part in the ushering in of the era of scientific botany.[22] But Nehemiah Grew, whom many credit with the 'discovery' of plant sex, occupies a much less ambiguous place in these histories, firmly associated with the side of botany and not philosophy.

Grew's various works on plants were collected together in 1682 under the title *The Anatomy of Plants. With an Idea of a Philosophical History of Plants, and Several Other Lectures Read before the Royal Society.*[23] This includes *An Idea of a Philosophical History of Plants* (1671),[24] *The Anatomy of Vegetables, Begun* (1672) and papers on the anatomy of roots

(1672/3), trunks (1673/4), leaves (1676), flowers (1676), fruits (1677) and seeds (1677). In *The Anatomy of Vegetables, Begun* Grew considers the use of the floral 'attire', which is his (collective) term for the structures in the flower that are now called the stamens and anthers and the style and stigma. As far as Grew is concerned, the attire is 'for us' partly ornamental (as the word 'attire' itself suggests), and serves to help us distinguish one plant from another, as well as serving as food for small animals. But these are the secondary uses of the attire, and what the primary or 'private' use may be, he says, 'I now determine not'.[25] Four years later, in a 1676 paper on the anatomy of flowers, Grew returns to the question of the primary use of the attire – that is, 'as hath respect to the *Plant* it self'. That there must be some such use, and an important one at that, he infers from the fact that all plants are 'attir'd', and conjectures that it performs some service to the seed. But what service?

In discourse hereof with our Learned *Savilian* Professor Sir *Thomas Millington*, he told, me, he conceived, That the *Attire* doth serve, as the *Male*, for the *Generation* of the *Seed*.

I immediately reply'd, That I was of the same Opinion; and gave him some reasons for it, and answered some *Objections*, which might oppose them. But withal, in regard to every *Plant* is *arrenothēlus* or *Male* and *Female*, that I was also of the Opinion, That it serveth for the *Separation* of some Parts, as well as the *Affusion* of others.[26]

These lines are, for many, the *locus classicus* of the beginnings of the theory of the sexuality of plants and are often treated as a turning point in botanical history. According to Ellison Hawks 'The recognition by Grew and [following him, John] Ray of the universality of sex among Flowering Plants was specially significant in that it was distinctly contrary to the long-received Aristotelian theory'.[27] Referring as well to the explanatory paragraphs that follow this first statement, Taiz and Taiz claim that 'Nehemiah Grew inaugurated the sexual theory by comparing stamens to penises, anthers to testicles, and pollen grains to sperm'.[28] Here, then, according to the received history of botany, several moves converge: (i) rejection of Aristotelianism, or rejection of the alleged Aristotelian rejection of the idea of male and female plants; (ii) the shift away from philosophy (theoretical speculation, metaphysics) to empirical observation and induction; and (iii) the shift of attention to the idea of the sexual *parts* of plants.

However, what Grew says in explaining his opinion – 'The sum therefore of my Thoughts concerning this *Matter*' – does not bear much of this out. According to Grew:

> And First, it seems, That the *Attire* serves to discharge some redundant *Part* of the *Sap*, as a *Work* preparatory to the *Generation* of the *Seed*. In particular, that as the *Foliature* serveth to carry off the *Volatile Saline Sulphur*: So the *Attire*, to minorate and adjust the *Aereal*; to the end, the *Seed* may become the more *Oyly*, and its *Principles*, the better fixed.[29]

This explanation draws on aspects of Grew's cosmo-physical and chemical theories, which he presented to the Royal Society in various papers and also expounded at length in his last long work, *Cosmologia Sacra*, 1701.[30] That the pollen was a kind of excremental discharge was an ancient idea, here combined with Grew's own theory that physical substances are compounds of immutable 'principles' (what others would call 'atoms'). As he explained in his 1674 paper to the Royal Society, principles (or atoms, or 'certain *Sorts* of *Atomes*, or... the *Simplest* of *Bodies*') are physically indivisible, impenetrable and immutable. Compound bodies are composed of many millions of these 'principles'. Grew says that principles are *propagated* like from like, which seems to refer to a kind of vegetative multiplication. Only compound bodies are *generated* from a conjunction of different types of principles. As all generated things are not the same – indeed they are very diverse – it must follow that there are numerously diverse 'principles'.[31] Mixture is to be understood in terms of the conjunction of different principles, the proportion of each and their location relative to each other. Because of their immutability and impenetrability, even the most perfect and subtle mixture, 'as in *Generation* it self', is in fact no more than contact.[32]

This physical theory is behind the idea that, by means of the attire, 'aery' compounds are discharged to leave the remaining physical mixture of the seed more oily, and the body more compact and close. Grew had observed that the body of the seed was composed of 'bladders' – what we now call 'cells' – that were smaller than the bladders of the other parts of the plant. He concludes that these bladders are smaller because, unlike the larger ones, they do not contain so much 'Aer', copiously mixed with sap (which mixture causes a quick fermentation and thus dilation and amplification of the bladders). This first explanation of the use of the attire aligns it explicitly with the *female* aspect of the plant: 'Wherefore, as the *Seed-Case* is the *Womb*; so the *Attire* (which always stands upon or round about it) and those *Parts* of the *Sap* herinto discharged; are, as it were, the *Menses* or *Flowers*, by which the *Sap* in the *Womb*, is duly qualified, for the approaching *Generation* of the *Seed*.'[33]

He continues (it is necessary to quote at length):

And as the young and early *Attire* before it opens, answers to the *Menses* in the *Femal*: so it is probable, that afterward when it opens or cracks, it performs the *Office* of the *Male*. This is hinted from the *Shape* of the *Parts*. For in the *Florid Attire*, the *Blade* doth not unaptly resemble a small *Penis*, with the *Sheath* upon it, as its *Praeputium*. And in the *Seed-like Attire*, the several *Thecae*, are like so many little *Testicles*. And the *Globulets* and other small *Particles* upon the *Blade* or *Penis*, and in the *Thecae*, are as the *Vegetable Sperme*. Which, so soon as the *Penis* is exerted, or the *Testicles* come to break, falls down upon the *Seed-Case* or *Womb*, and so Touches it with a *Prolifick* Virtue... That the same *Plant* is both Male and Female, may the rather be believed, in that *Snails*, and some other *Animals*, are such. And the *Parts* which imitate the *Menses*, and the *Sperm*, are not precisely the same: the former being the External *Parts* of the *Attire*, and the *Sap*, which feeds them; the latter, the small *Particles* or moyst *Powder* which the External inclose.

And that these *Particles*, only by falling on the *Uterus*, should communicate it to the *Sap* therein, a *Prolifick Virtue*, it may seem more credible, from the manner wherein *Coition* is made by some *Animals*; as by many *Birds*, where there is no *Intromission*, but only an *Adosculation* of *Parts*... Nor doth perhaps the *Semen* it self [enter]: or if it doth, it can by no means be thought, bodily or as to its gross *Substance*, to enter the *Membranes*, in which every *Conception*, or the *Liquor* intended for it, before any *Coition*, is involved; but only some subtle and *vivifick Effluvia*, to which the visible *Body* of the *Semen*, is but a *Vehicle*... [But] If any one shall require the Similitude to hold in every Thing; he would not have a *Plant* to resemble, but to be, an *Animal*.[34]

Several things here need to be considered: Grew's first claim that 'the *Attire* doth serve, as the *Male*, for the *Generation* of the *Seed*'; the comparison in the 'florid attire' of the plant 'blade' with the animal penis and the 'sheath' with the prepuce; the comparison in the 'seed-like attire' of the thecae with the testicles and the 'globulets' with sperm; the hermaphroditic nature of the attire and thus of plants; and the idea of a 'prolifick virtue' and 'vivifick effluvia'. A close consideration of these things in the context of Grew's philosophy, which does much to explain them, shows that the interpretations of Grew's remarks in the traditional histories of botany tend to presume that Grew saw what we think of as the sexual parts of the flower, when his view was somewhat more complicated and ambiguous.

First, by 'florid attire' Grew means what we would now call the florets of composite flowers, like the marigold, one of Grew's examples;[35] the 'seed-like' or 'seminal' attire refers to the stamens and anthers of simple flowers, like the tulip (again, one of Grew's own examples). In *The Anatomy of Plants, Begun* Grew describes the florid attire as comprising usually three pieces or 'suits' – the outer floret, within this the sheath and the innermost part, the blade. Grew refers us to an illustration (see Figure 3.1), which along with this 'Explication of the Tables', helps us to understand to which parts he is referring.[36] His 'Peruvian Starwort' is obviously an example of the Asteraceae family (also called Compositae), which includes daisies and sunflowers. What he calls the 'sheath' (in his figure labeled 13.c) is what is now identified as a ring of fused stamens. This sheath divides at the top, he says slightly later, and from within this third and innermost part of the 'suit' emerges the blade (in his figure labeled 13.d). This 'blade' is what is now identified as the pistil (style and stigma): that is, the 'female' part of the plant.

When Grew recounts that Millington 'told, me, he conceived, That the *Attire* doth serve, as the *Male*, for the *Generation* of the *Seed*', this is usually interpreted as the '*stamens-as-penis* hypothesis', as Taiz and Taiz put it.[37] But it is the 'blade' (the 'female' style and stigma) that Grew explicitly compares to the penis.[38] As Grew says, this is suggested by the shape of the parts; and, indeed, the style/stigma is very often the most phallic-looking part of the flower: consider the lilies. However, what is important is not so much the correct identification of a part as the identification of *functions* – and specifically the male 'office' – opening the way for a new conception of male and female in plants. But what is this?

Grew describes the seed-like or seminiform attire as made up of two parts, 'chives' and 'semets', there usually being two semets, filled with powder, on each chive.[39] The tulip is used as an example here. The 'semets' are what are now called the anthers, and they are said to be the 'thecae' or 'cases' 'of a great many extream small Particles'.[40] In his later paper on the anatomy of flowers Grew also identifies another distinct part

FIGURE 3.1 Grew, *Anatomy of Plants Begun*, Tab. IV. 'a' is a floret, with sheath and blade; 'c' is the sheath removed; 'd' is the blade removed. By permission of the Linnean Society of London.

of the seed-like attire, 'like a little *Columna* or *Pinacle*, which stands on the *Top* of the *Uterus* or *Seed-Case*' – presumably the style/stigma. When he thus compares the 'thecae' of the seed-like attire to testicles, he does get the sex right, from the standpoint of modern botany. So too does he when he compares the globulets of the thecae to sperm – but he also claims that the same 'vegetable sperm' is produced by the blade (i.e. the 'female' part, in modern terms). In Grew's imagination, then, the blade/stigma is compared to the penis, and the semets/anthers to the testicles. In using the common language of the 'uterus' he distinguishes (correctly, from the modern standpoint) a female part that receives the sperm, but on the whole his identification of the sex of parts is a hit-and-miss affair and in certain respects he is simply wrong.[41] Again, however, what is most important is the general framing point: that in plant generation, as in the animal, we may distinguish between the male and female functions or offices. The hermaphroditism of the plant is the coincidence of the parts for the performance of these two offices in the same flower. And the 'male office' concerns the transmission of what Grew calls the '*Prolifick* Virtue' or '*vivifick Effluvia*', an understanding of which requires us to consult his philosophical, as well as his botanical works.

The male 'office' and the '*Prolifick* Virtue'

Grew's philosophy is expounded in his last work, *Cosmologia Sacra* (1701). Here, in the first chapter of the Second Book, 'Of Life', Grew aims to prove 'That there is a Vital Substance in Nature, distinct from a Body'.[42] Grew argues against what he takes to be Descartes' and Thomas Willis's view that the life of an animal could be in a 'very Subtile, Aerial, Etherial, or Igneous Fluid' in the body, because no matter how 'Subtilized' such a fluid might be it would still be composed of the atoms which are 'no more, than Parts of the Common Stock of Body', and body cannot be vital.[43] Nor, according to Grew, does the fact of the organization of a body make it vital (alive), as organization requires only bulk, figure and mixture, or that the parts of the body 'be fitly Cized, Shaped, and set together'. If these were sufficient to make a body vital, then all bodies would be alive, as all have bulk, figure and mixture.[44] Finally, Grew also denies that motion produces or is a mark of life, as this would mean that the quicker a body, the more life it had (so that light was the most alive of all things), and that engines produced by human art would become vital in being moved.[45] This being so, it must be the case, according to Grew, that 'we must allow the being of a Substantial Principle, distinct from Body, as the proper and immediate Subject of Life… the proper and immediate Subject, not of one only, but of any Species of Life'.[46] This 'subject' is 'something, which is Substantial, yet Incorporeal',[47] but Grew refrains from calling it 'soul'. Just as motion is the immediate 'adjunct' (i.e. proper but not essential accident) of body, so life is the immediate adjunct of the vital substance: 'What therefore Motion is, to all Bodies; that Life is, *Suo modo*, to all the Species of Vital Substance. By mediation of which Two Adjuncts [motion and life], there is an easie Commerce between Things Corporeal and Incorporeal.'[48] Body (or matter) and motion are the two visible parts of the universe,[49] which allows us to infer that 'life' and 'vital substance' are invisible.

Grew's argument is not easy to follow, because the distinction between the proper subject of life (immaterial substance) and life itself is not consistently made and Grew equivocates between different meanings of 'principle' – both 'atom', as in the earlier work, and 'power'. Grew claims that there is a 'universal stock' of matter and motion which can be neither increased nor diminished, but only transferred. Any regularity in the transfer of matter and motion – including, remarkably, any regularity in the movement, size or figure of any corporeal body, which must include mineral bodies – cannot be due to body itself, but needs for its explanation 'such a Vital Principle, as aforesaid, to determine it. Of which Principles, we must then allow a Stock, answerable to the Corporeal, as one Moiety of the Universe'.[50] He continues:

On the Directive Power of the former [the vital principle], and the Regularity of the latter [motion], whereby it is capable of Direction; depends the Generation of all Bodies. The said Power, being one and the same Vegetable Life, infused into all the Parts of Corporeal Nature; but more remarquably into Plants and Animals.[51]

Grew earlier claimed that the several species of life are reducible to three, '*viz.* Vegetable Life, Sense, and Thought'.[52] The principles of vegetable life direct the motions of nutrition, augmentation and generation in all species of life, such that the animal too (including the human animal) has its vegetable life,[53] but the other species of life – sense and mind – have their own, distinct vital principles too.[54] As with Aristotle, the commonality of vegetable life is the foundation or condition of possibility of the plant-animal analogies concerning the parts and functions of nutrition and generation – the fact that they share this 'species of life', as Grew calls it. This means that the analogies are not only the result of the observation of visible similarities but the result of inductive reasoning on the basis of a developed philosophical theory – just as in Aristotle.

Further, Grew understands the generation of plants (as of all living things) in terms reminiscent of the generation of animals in Aristotle and the Aristotelian tradition. In *Cosmologia Sacra*, referring back to his botanical works, Grew writes that 'In the consideration of Plants, I have set down the Method of Generation, step by step, as far as the Regularity of Principles will go. But for the Performance of this Work, a Vital or Directive Principle, seemeth of necessity to be assistant to the Corporeal'.[55] This, it seems, must be what he means when he refers, in his discussion of the 'use' of the attire, to the '*Prolifick* Virtue' or '*vivifick Effluvia*' of the particles (i.e. pollen) which he believes to come from both thecae (anthers) and blade (style/stigma). There he wrote, recall, that these particles 'only by falling on the *Uterus*, should communicate it to the *Sap* therein, a *Prolifick* Virtue' but, as in some animals, by adosculation, not intromission.

> Nor doth perhaps the *Semen* it self [enter]: or if it doth, it can by no means be thought, bodily or as to its gross *Substance*, to enter the *Membranes*, in which every *Conception*, or the *Liquor* intended for it, before any *Coition*, is involved; but only some subtle and *vivifick Effluvia*, to which the visible *Body* of the *Semen*, is but a *Vehicle*.[56]

We can now understand this to refer to the invisible, vital principle that is *carried by* the body of the sperm but is *not identical with* its material principles/atoms, just as, in Aristotle, the male imparts the sensitive soul or motion through a power that is not identical with its material vehicle. Although this tends now to be overlooked in discussions of Grew's botany, the idea of this invisible, non-corporeal vital principle is as much a part of his remarks on plant sexuality as the identification of the pollen as the (male) sperm (albeit he also thought the style contributed pollen). To say that the attire 'performs the *Office* of the *Male*' means both that it sheds the pollen (semen) on to the uterus and, *just as importantly*, that it provides the vital principle. Indeed, this may be what *defines* this use of the attire as male. Thus, what was an obstacle to the application of an Aristotelian view of the distinction between male and female to plants (the idea that the male provided the sensitive soul) is overcome with the new idea that the male provides, more generally, a non-material vivifying principle in all vegetable life (plant, animal and human), or that the male provides the life itself.

It is also often overlooked that John Ray (*Historia Plantarum Generalis*, 1686) paraphrases Grew's opinion on this in full, mentioning not only that he sees the globules of the stamen as the fecundating male sperm (the part that the received history of botany repeatedly cites) but also that the sperm does not actually penetrate the uterus or the seed (as it does not penetrate the ovary of an animal), it being sufficient for an 'exhalation' or 'subtle effluvium' to fecundate the egg or to vivify the embryo in the uterus.[57] When Ray thus says that Grew's theses require confirmation (Ray admits them as only probable) it is this whole account of plant generation that is at issue. Finally, in 1694, in the Preface to his *Stirpium Europaearum*, when Ray affirms aspects of Grew's theses more forcefully, he says that the pollen of the plant is analogous to the sperm of animals and is endowed with a life-giving power (*vi prolifica dotatum*), which, again, seems to be the same as Grew's 'prolifick virtue'.[58]

From a modern standpoint it may be tempting to treat Grew's and Ray's talk of prolific virtues and vivifying effluvia as extraneous metaphors, in order that the literal identification of the male and female parts of the plant may stand out in the history of the theory of plant sexuality. But this misrepresents the theoretical role that they play in Grew's and Ray's work in thinking of the possibility of male and female parts in the generation of plants. It selectively reads modern botanical knowledge back into them, in order to find itself reflected there. Zirkle, for example, claims that already in 1672 (in *The Anatomy of Plants, Begun*) 'it is practically certain that Grew knew the primary function of the pollen', that he definitely did in 1676 (in the paper on the anatomy of flowers), and because of this 'became aware that the flowering plants reproduced sexually'.[59] Ellison Hawks writes that, by 1690 Ray was 'fully convinced that the stamens were male organs'.[60] But what did it mean, for Grew and Ray, to think of a part and its function as 'male'?

In both Grew and Ray sex in plants was presented and understood in terms of contemporary (mainly Aristotelian) theories of *animal* generation in which the male played the enlivening, animating or inspiriting role. Grew's position, more particularly, follows from his general conception of life, which includes plants, and requires a 'Vital or Directive Principle' that is associated with the male office. In beginning to think of plants in sexed terms Grew is thus, via his conception of life, applying the existing theories of sexed (animal) generation consistently. That is, his conception of life *requires* for the plant the same vivifying role that the male plays in the generation of animals. The characterization of the attire and the function of the pollen as male cannot be understood simply in terms of the need to account for what is observed; it is also the result of the need, determined by Grew's conception of life, *to find a part* of the plant that can play the male, *vivifying* role. That is, it is not just that it is recognized that a part (the attire and its pollen) has a role to play and that it is thus called 'male' but that the Aristotelian presumptions of the more general theory of generation and Grew's conception of life *require* there to be a male part, which is subsequently located in the attire and its pollen.

Histories of philosophy have forgotten Grew, and histories of botany are embarrassed by his philosophy.[61] But the relationship between philosophy and botany in his work has interesting consequences for the way that we think about the history of both disciplines. No doubt Grew did observe the parts of plants more carefully and in greater details than most of his predecessors. But observation and induction are not opposed to philosophy and deduction or speculation in his botanical work. The advances he makes in the theory of the sex of plants are not in opposition to a metaphysical conception of 'male' and 'female', but rather driven by it. His botanical work lives and thrives on a philosophical basis. At the same time, philosophy proceeds through botany in Grew's work, as the need to account for the life and generation of plants in part motivates the philosophical reflection on male (especially) and female. Grew's work is a joint venture of both philosophy and botany, and this is nowhere more clearly seen than in precisely that part for which he is principally remembered: his contribution to the history of the theory of plant sexuality.

4 THE CARNIVAL OF PLANT SEX: FROM ANALOGY TO IDENTITY

We saw in the previous chapter that with Zalužanský, Grew and Ray the history of the discovery of plant sexuality is an episode in plant philosophy, quite as much as it is a defining chapter in the history of botany. Although we can emphasize one or the other, botany or philosophy, in the texts and arguments of these thinkers themselves the two are inseparably entwined.

In addition to Zalužanský and Grew, the other main candidate for the distinction of having discovered plant sexuality is Rudolf Jakob Camerarius. He, too, continued to treat Aristotle's philosophy as a scientifically authoritative source, but his use of experiment to prove the necessity of both 'male' and 'female' in plant fertilization marks a significant shift in the discourse of plant sexuality. However, Camerarius's contribution introduces a paradox into the history of scientific botany – a paradox of some interest to plant philosophy. In the literature on the use of analogy (or analogical reasoning) in the sciences there is a broad consensus that, while often initially useful, the danger inherent in all use of analogical reasoning and modelling is the tendency for the recognition of similarities between different things to slip into their being treated as identities. But when it comes to the history of the theory of the sexuality of plants, the great scientific step is, on the contrary, taken to be the move initiated in Camerarius's work *from analogy to identity* – to the recognition of the basic, *literal* identity of the sexual function in animals and plants. What is the significance of this peculiarity in the history of botany?

In this chapter we will look at this move from analogy to identity in Camerarius's *De Sexu Plantarum Epistola* (*Letter on the Sex of Plants*, 1694) and at the popularization of the idea of plant sexuality in Sébastien Vaillant's *Discours sur la structure des fleurs* (1718) and various of Linnaeus's works. In so doing, we come across a second paradox. Why is it that in these texts that are supposed to popularize the literal truth of plant sex the discussion of plant sexuality becomes so egregiously literary and metaphorical? Why do we find more extended and tenuous analogies between the details of the 'nuptials' of animals and plants just at that moment that the literal truth of plant sex is being affirmed? And what implications can plant philosophy draw from these paradoxes? In this chapter we argue that the works examined here point not to the existence of separate scientific and literary discourses on the sexes of plants but to the ineliminable ambiguity between the literal and the metaphorical when it comes to vegetal sex, an effect of the seemingly constitutive equivocation in 'male' and 'female' between 'function', 'organ' and 'individual'.

Rudolf Jakob Camerarius: 'Actually and literally as such'

As Sachs notes, the style of Camerarius's *Letter on the Sex of Plants* 'is thoroughly that of modern natural science'.[1] For Sachs this is the result of Camerarius's concentration on the accurate description of the parts of the flower, the reporting of his experiments and his critical, analytical approach to the existing literature. As Aristotle's influence was still strong Camerarius felt duty bound, according to Sachs,[2] to show that his claims did not contradict Aristotle's but the move from philosophy to scientific botany is decisive.

Sachs overstates the extent of Camerarius's distance from Aristotle.[3] But what is more important is the role ascribed to Camerarius in the move from the understanding of plant sex as merely *analogous* to that of animals to an acceptance of its *literal* truth. As we have said, this is a paradoxical move in the context of the legitimate role usually ascribed to analogy in the sciences, where the move from similarity to identity – here, the recognition of the *literal* truth that plants are sexed – is usually characterized as misuse.

As a form of inductive reasoning, analogy allows researchers to identify, interpret and speculate about novel phenomena by virtue of their similarity (of feature, function, relation and so on) with what is already known or understood (or at least thought to be so), thereby allowing for the extension of categories, concepts and causal explanations into new areas. It may be the basis for testable predictions about new or as-yet poorly understood phenomena. Even if the logical validity of inferences from similarity is philosophically problematic and analogy cannot serve as the basis of proof, it is undeniable that analogy has served, historically, as a useful scientific heuristic (for discovery, rather than justification).[4]

However, analogy's condition of possibility – the positing of a similarity – is also its lure, for two main reasons. First, the analogy does not merely point out a similarity, it also constitutes it, at least in part. It allows us to see the similarity, the observation of which is supposed to be the basis of the analogy.[5] Second, if the similarity is compelling (which it is required to be for a successful analogy) there is always the danger that what is essentially a metaphorical relation will ossify into the presumption of a literal identity between the terms, or a supposedly literal application of what was originally a metaphorical extension of a concept to a new field. This is what Max Black calls the 'metaphysical' use of analogical extension.[6]

As many have pointed out, analogy is most frequently found in the biological sciences, which would also make biology particularly prone to the dangers associated with this form of reasoning. Agnes Arber, indeed, suggests:

> It may not be going too far to say that the history of intellectual movements in biology has been largely the history of analogies which were valuable within their true limits, but which had an irresistible tendency to exceed these limits, and thus to pretend to the status of identities. When the hollowness of this pretension comes to be recognized, the analogy is liable to be cast on one side as useless.[7]

This is especially true of botany and plant philosophy, given the role within them of analogies with animals. Bringing these points together, Arber writes:

> When we say that an analogy is pressed too far, we generally mean that it is, in reality, treated as if, rather than an analogy, it were an identity. This particular misuse of analogy is one to which biological thought is particularly liable. It beset the comparison between animals and plants, which was basic in Greek biology – a comparison which would have been helpful if it had been kept in its place, but which was carried so far that it led to serious misunderstandings of vegetable function and structure, which lasted into the seventeenth century and

later. Nehemiah Grew had detected this fallacy when he wrote 'If any one shall require the Similitude to hold in every Thing; he would not have a *Plant* to resemble, but to be, an *Animal*.'[8]

Except, it would seem, when the analogy concerns plant sex, where, on the contrary, the move to identity is taken to be the decisive scientific move, away from philosophy.

This move can be mapped within Camerarius's *Letter on the Sex of Plants*. In it, Camerarius distinguishes between three types of plants: those with stamen and style in the same flower (which we would now call hermaphroditic), those with stamen and style on different flowers of the same plant (now called monoecious) and those with stamen and style distributed across different individuals (now called dioecious). In whichever way the different parts are distributed, experiments have shown, Camerarius says, that without both anthers and style no seed or fruit will form.[9] The point of the *Letter* is then to ask, what, on this basis, can be determined about the sexes of plants?[10] To answer this Camararius turns to the animal kingdom where, he says, according to unanimous agreement, a sexual difference everywhere obtains, and it is easy to know the difference between male and female organs and their function. Leaving aside the difficult question of spontaneous generation, it is agreed that in the animal kingdom the male seed is necessary for generation; that certain parts are 'destined' for the formation and protection of this 'seed'; and, at the right time, a properly prepared and 'spiritual' fluid is evacuated. The female uterus and ovaries (sometimes also called testicles) are destined for the conception and perfection of the foetus. This is the basis for what J. G. Gmelin, in his Introduction to an edition of Camerarius's letter in 1749, calls a 'maximal analogy' with animals.[11] The analogy between male plant 'seed' (what we now call 'pollen') and animal semen justifies, according to Camerarius, the pollen and the anthers being assigned the 'noble name and the function of the genitals of the male sex'. This male seed is 'the most subtle part of the plant', secreted 'when properly percolated and refined'. Correspondingly, the seed pod with its 'plumula' or its style is the female genital organ, receiving and protecting the male seed, performing the 'maternal' office.[12]

As Sachs says, Camerarius is primarily interested in establishing the fact of plant sexuality, not in developing a theory of generation.[13] Nevertheless, he does briefly discuss controversies in the theory of the generation of animals. Ostensibly, he remains agnostic with regard to both ovist preformationism (which posits that the egg containing the new animal is animated by the influence of the male) and spermaticism (which posits that the tiny new animal is contained within the male seed).[14] He says that the fact that this controversy is not settled in the animal kingdom, where things are so much better known and everyone accepts the fact of sexuality, means that we should not expect to understand the precise relation between male and female in plants. However, Camerarius does say that it is not possible to see a seed in either the anthers or the pollen ball, which speak against spermaticism. He also says that it is not possible to see a seed in the unfertilized ovary, and that the doctrine of ovist preformationism makes it impossible to ascribe any role to the male in plant generation, and thus impossible to call any plant or part of the plant male. As everything that follows in his *Letter on the Sex of Plants* affirms the necessity of the male element (the pollen) in plant generation he also effectively criticizes ovism. Camerarius discusses various accounts of the generation of frogs and fish in Aristotle, Theophrastus, Scaliger and Harvey (where the male sprinkles his milt on the female's eggs, rather than introjecting it into her) and he approvingly quotes 'Aristotle's' claim that the male contribution is more qualitative than quantitative, and Ray's claim (which we cited in the previous chapter) that the semen does not enter the ovary but rather fecundates it with a vivifying '*effluvia subtilia*'.[15] According to Camerarius, just as the superficial contact of the fish milt on the egg is sufficient for fecundation, so is the superficial dusting of the pollen on the style, the wind performing for the pollen the role of the water for the fish milt. Just as the phenomenon of the avian wind egg shows that the male is necessary for fertilization, so does the sterile plant. For granted that heat and moisture are necessary for the germination of seeds, these are not themselves the '*vis genitalis*'.[16]

The consensus on this point between Zalužanský, Grew, Ray and Camerarius is striking. Although the traditional histories of botany pass over this, at this crucial moment at the end of the seventeenth century and the very beginning of the eighteenth, the 'discovery' of the sexes of plants in something approximating the modern view of things is not just the recognition of the male and female parts of the flower but also the affirmation of a quasi-Aristotelian view of the male office. The male office in plant fertilization can be affirmed even in the absence of evidence of 'intromission' (the penetration of the pollen into the ovary) because of its conception as a subtle, spiritual, vivifying effluvium or 'prolifick virtue'. This ultimately philosophical conception of the male office is an essential part of these arguments for the sexuality of plants, not merely its old-fashioned garb.[17] Camerarius's contribution was to offer experimental proof for the necessity for the male office, which proof (if it was accepted; not everyone did accept that the sexes of plants had been proven[18]) was exceptionally important in promoting the idea of the sexes of plants.

But Camerarius was also at pains to point out the meaning of the idea of the sexes of plants or to specify in what sense we should understand them to be sexed. Although he and others sometimes spoke of the male 'seed' of the plant, the proper comparison, according to Camerarius (and many others), was between the fertilized seed of the plant (the result of the mixture between male and female) and the fertilized egg or embryo.[19] As we saw in Chapter 2, the seed-egg comparison was already made in Aristotle and the Aristotelian botanical tradition. For Aristotle the male and female principles are, as it were, from the beginning already mingled in the plant seed. But once having identified the different *parts* of the flower (anthers and style) as male and female, Camerarius is able to claim that male and female are separate in the plant (even in hermaphroditic plants) *just as they are in animals*. It is this *separation* that is explicitly at issue when Camerarius asserts the literal truth of plant sex. For, he says,

> why should they [plants with male and female in separate flowers or on separate plants] not also exercise these functions [male and female] and be designated with these names? They [plants] behave indeed to each other as male and female, and are otherwise not different from one another. They are thus distinguished with respect to sex, and this is not to be misunderstood as it is ordinarily done, as a sort of comparison, analogy, or figure of speech, but is to be taken actually and literally as such.[20]

Commentators affirm this move from analogy to identity (to the *literal* idea of plant sex) as a major step in the history of the theory of the sexuality of plants.[21] So here, it seems, we have a unique counterexample to the idea, held quite generally in discussions of analogy in the history of the sciences, that the move from analogy to identity is an abuse of analogical reasoning. This exceptional move can perhaps be understood if we examine what, exactly, is being identified in animal and plant. Camerarius says that the similarity of function between animal semen and pollen allows us to say that the anthers and stamens are the male sexual parts and accordingly the seed pod and its plumula or style correspond to the female sexual parts (or the maternal function). Any introductory book on botany today will (*mutatis mutandis*) agree with this, but it is not exactly the point at which Camerarius proposes the *literal* identity between aspects of the animal and the plant. Indeed, immediately after this he implicitly notes a significant *difference* between the sexual parts of animals and plants, namely that the organs of generation are ephemeral in plants, which are forced to produce these organs and the seeds within anew each year.[22] Further, a plant does not have just one uterus but each year produces many. Each such uterus is not 'perpetually thriving' but only fertile for a short time every year such that one can ask, as Theophrastus did, says Camerarius, whether each time it is the same or a different plant. These points stress the grounds for a disanalogy between the 'sexual organs' of animals and plants, a disanalogy that might lead to something more akin to a plant philosophical insistence on the specificity of the plant.

In the following paragraphs Camerarius discusses Aristotle's view of animal generation and focuses particularly on the movement of animals towards each other in copulation. At the same time as male animals, he says, begin to form their seed, female animals become ready for conception and they are driven to unite and copulate so intimately that we say that they (male and female) become one flesh. According to Camerarius, we see this same movement in plants, as the anther and the style come together. In plants with male and female in the same flower they stay together like this (he claims) until the flower drops. Camerarius then notes how Aristotle remarked this union of male and female in the plant and how animals, like plants, appear in their undivided nature in copulation, where they strive to become one. As plants have no other aim than generation, male and female remain mixed in them, whereas animals are mixed only at the time of copulation so have separate sexes. Camerarius thus interprets Aristotle's claim that the metaphysical principles of male and female are already mixed in plants to be saying that the male and female parts are found together and mixed in a kind of perpetual copulation, in hermaphroditic plants at least. Noting that most plants are hermaphroditic, Camerarius presumes that these are all self-fertilizing and the impression is given that, for him, this hermaphroditic state represents the true nature of the plant. This being so, what needs to be argued is that other kinds of plants – those with 'male' and 'female' separated in different flowers on the same plant or in different plants – can also genuinely be called 'male' and 'female'. Camerarius does not reject Aristotle's view that male and female are already mixed in plants; rather, he affirms the sense in which this is true of hermaphroditic flowers and justifies the extension of 'male' and 'female' to monoecious and dioecious plants where these functions are separated (in different flowers or on different plants). This is affirmed against the kind of metaphorical view described by Theophrastus, for example, where individual plants are called 'male' and 'female' on the basis of the presumed masculine or feminine qualities of the bark, wood and so on.

How is Camerarius's affirmation related to the modern conception of the literal truth of plant sex? Histories like those of Sachs, Morton and Taiz and Taiz are written from the standpoint of their respective presents, with a modern biological definition of sexuality in mind. In his *Lehrbuch der Botanik* (*Textbook of Botany*), first published in 1868, Sachs included a chapter on sexuality and defined it thus: 'Sexuality is based on the fact that in the course of plant development two kinds of cells are produced. In isolation, these cells are not able to develop; only when these cells combine, a product is created that has the potential for development'.[23] As Ulrich Kutschera and Karl J. Niklas further specify: 'Thus, Sachs defined sexuality as the fusion of male and female gametes (spermatozoids and egg cells) and explicitly called this merger of male/female gametes the Geschlechtsact (sexual unification)'.[24] In this definition 'male' and 'female' refer either to the gametes themselves or, by implication, to the bearers of gametes. As we shall see in more detail in Chapter 5, the bearers of gametes in vegetal life are the gametophyte generations, which in most plants are not the visible, well-known forms and flowers of adult plants but are cells within the pollen grains and the embryo sac in the ovary and which are separate plants. As Lincoln Taiz's textbook *Plant Physiology and Development* thus says, 'Strictly speaking, the flower itself is not a sexual structure'.[25] James D. Mauseth's textbook of botany similarly points out that '[s]tamens are frequently referred to as the "male" part of the flower because they produce pollen, but technically, they are not male because the flower, being part of the sporophyte, produces spores, not gametes. Only gametes and gametophytes have sex'.[26] Stamens in hermaphroditic plants, staminate flowers in dioecious plants and monoecious plants with staminate flowers are thus not *literally* sexed or *literally* male, according to the modern definitions of male and female, but only conventionally so designated. (We discuss this at greater length in Chapter 5.)

Of course, Camerarius knew none of this. For him the male was that part of the flower, or that flower or that kind of plant that animated the seed via the spiritual exhalation of the pollen, and the female was that part of the flower, or that flower or that kind of plant that was the provider of the egg and, once it was fecundated, its maternal nurse and incubator. His claim is that these are literally male and female, in the same way that male and female animals and their sexual organs are literally male and female. It is a claim about sexed individuals and their organs, based on an animal model. It is possible to translate Camerarius's claims into the formulations of modern scientific understanding, but it remains the case that what he asserted as literally true, and for which he is afforded so much credit in histories of botany, *is not* literally true. It is only 'in a manner of speaking' that the stamens and staminate flowers and staminate plants are male. But the point here is not to argue that Camerarius was wrong. It is that the example of Camerarius shows that it is very difficult to separate out what is literally the case from our manner of speaking about it when it comes to the sexuality of plants, and that this has to do with the multiple senses of 'male' and 'female' (as principles, individuals and organs) and the conflation of these.

The nuptials of plants: Sébastian Vaillant and Carl Linnaeus

After Camerarius, various others attempted to confirm the same experimental finding that the pollen is necessary to the fertilization of the seed. Many of these writers and botanists are now little known outside of specialized histories of botany. For example, in France in 1711, Claude Joseph Geoffroy reported experiments with corn and affirmed that the male anthers 'must be regarded as the principal cause of the fecundity of plants'.[27] In England Richard Bradley's *New Improvements of Planting and Gardening* (1717), drawing extensively on 'parallels' between animals and plants, accepted the idea of male and female in plants and describes experiments with tulips and hazel and apple trees.[28] In 1735 James Logan reported to the Royal Society from Pennsylvania his experiments with maize.[29] Amongst others who discussed the idea of sexuality in plants and reported on their observations of plant structure, the major issue of contention, once plant sexuality in some sense was basically accepted, was whether or not the pollen actually penetrated the ovary. In 1703 Samuel Morland, referring to Grew rather than to Camerarius, contested the idea that the pollen need only touch the outside of the 'uterus', suggesting instead – following Antonie de Leeuwenhoek's animaculist model – that the pollen is a '*Congeries* of Seminal Plants' and that the style is 'a Tube design'd to convey these Seminal Plants into their Nest in the *Ova*'.[30] Patrick Blair discussed this controversy in 1720, in his *Botanick Essays*. Referring to Grew, Ray, Camerarius, Leeuwenhoek, Geoffroy, Richard Bradley and Samuel Morland, Blair finds that they have diametrically opposed opinions. Either 'the Farina falling upon the Pistillum Vasculum Seminale or Semen, impregnates the Seed it self, and there actuate upon the gross Particles previously in the Seed-Case or Uterus: Or it is a Congeries of Seminal Plants, one of which must enter the Vasculum Seminale, and there become the Semen, as Mr Morland and his Adherents would have it'. Blair, having examined many plants, writes that he 'cannot find the least sign of Probability for Mr Morland's Opinion' and he affirms instead Grew's position.[31]

Within this history, until the middle of the eighteenth century, Sachs and others distinguish between those who made genuine scientific contributions to the theory of the sexuality of plants (notably Camerarius and those who repeated the same kind of experiments to show *that* the pollen was the fertilizing agent, and thinkers like Morland and Blair who tried to understand *how* it was) and those who merely popularized the theory.[32] But because the popularizers were, precisely, popular, it is their names that are better known and have come to be associated with the idea of the sexuality of plants (indeed, they were earlier sometimes credited with having discovered it). Chief among the popularizers are Sébastien Vaillant and Linnaeus. But what, exactly, in the theory was being popularized? Was it the literal truth of plant sex?

Both Vaillant and Linnaeus mention various experiments that offer proof of the necessity of the male function in plant fertilization, and it is clear that, for both, this was no longer an open question. Vaillant's *Discourse on the Structure of Flowers*, published in 1718, was the text of a lecture (the first in the annual course on botany) given on the occasion of the opening of the Royal Garden of Paris in its new location in June 1717. It is an early example of a popular scientific genre. Vaillant's scientific aim in the lecture is to affirm that flowers are the sexual organs of a plant, to establish a terminology for these sexual organs, to affirm Grew's view of the male office and to describe the relation between numbers of stamens and petals in flowers as a means of identification, and thus classification, of flowering plants. Vaillant tends to refer only obliquely to the various botanists upon whose works he draws (including Malpighi, Grew, Ray, Joseph Pitton de Tournefort and Geoffroy) or whose ideas he criticizes, even ridicules (principally Geoffroy but also implicitly Tournefort).[33] He takes for granted Cesalpino's point that the reproductive organs are the most essential part of the plant and that these parts can be understood according to the animal model. In his discussion of the function of the pollen he claims (like Grew) that it is not the material body of the pollen but the 'vapour, or volatile spirit' which fecundates or 'vivifies, animates' the plant 'egg' just as the gross matter of animal semen is only the vehicle of the volatile animal spirit: 'Because nature acts always according to uniform Laws, one must conclude that what happens on this occasion in animals, must happen in the same way in Vegetables.'[34]

It is not clear if Vaillant knew Camerarius's work first-hand, but he writes of the sexuality of plants as an established fact and presumes, a priori, the universality of basic function in living beings (a condition of possibility for animal-plant analogies). Despite no explicit reference to Camerarius, Vaillant's lecture can read as assuming the literal truth of plant sex in some sense. Nevertheless, the innovation in Vaillant's presentation of the sexuality of plants, especially in the opening section of the lecture, is an extended, ribald analogy with animal sex and with human copulation and nuptials. This is probably the first example, then, of what will henceforth be an insistent feature of the discussion of sexuality in the plant advocacy literature: a somewhat comedic, and even puerile, nudge-nudge wink-wink tone and the personification of plants as beings with sex lives and love lives.

Vaillant says that flowers can only be understood as 'the organs which constitute the different sexes of plants', which are principally two: the stamens and the ovaries.[35] The stamens are the masculine organs, the most noble part of the flower 'because they correspond to those which in male animals serve for the multiplication of the species; these, I say, are composed of testicles [testes] and penis [queûes], or if one prefers to stick with the ordinary terms, anthers [sommets] and stamens [filets]'. Bernasconi and Taiz translate Vaillant's 'queûe' as 'tail',[36] but as the word was being used to refer to the penis by the eighteenth century, and given the tone of Vaillant's lecture, it is not unlikely that his audience heard 'penis' (or, rather, a slang word for 'penis'). This is also suggested by Vaillant's claim that the tail/penis/filament is properly only the sheath for the spermatic vessels.[37]

Vaillant justifies the equation of anthers with testicles not just because they often have the same 'figure' (by which he could mean the same number or the same shape) but also because they perform the same office, being two 'cartridges' (*cartouches*) or membranous capsules which essentially have two compartments full of 'dust'.[38] He describes how the petals of flowers, especially when in bud, serve to protect the sexual organs of the plant. In hermaphroditic plants the bud hides the generative organs of the plant completely and can thus be seen as the 'nuptial bed, since it is usually only after they [the organs of generation] have consummated their marriage that they are allowed to show themselves to each other'.[39] Where separate male and female flowers are found on the same plant, things must obviously be different. There

> the tension or the swelling of the masculine organs happens so rapidly that the lobes of the bud, giving way to their impetuosity, open up with surprising speed. At this moment, these ardent organs [*ces fougueux*], which appear to seek only to satisfy their violent transports, not earlier feeling free, [now] suddenly discharge all about a tornado of spreading dust which carries fecundity everywhere.[40]

The stamens of hermaphroditic flowers do not act with the same alacrity or vigour but, Vaillant writes, we can presume that the slower they are, the longer their 'innocent pleasures' endure. Vaillant also speaks of the 'frolics' of the generative organs of plants, and their 'amorous exercises',[41] but in fact only the male organ is identified with the organ of sexual pleasure in the animal, and the frolics are all his. For Vaillant the female sexual organ of the plant is the ovary, which he says others have wrongly called the pistil or calyx. Here the seeds (which are true eggs) are nourished until they mature. Vaillant divides the ovary into its 'belly' (*panse*) and 'trunk' (*trompe*).[42] The latter refers to the style, but Vaillant explicitly compares it to the Fallopian tubes (*trompes de Fallope*) because 'they transmit to the little eggs, not the grains of dust themselves which ejaculate on them… but only the vapour, or volatile spirit coming from the grains of dust, which will fertilize the eggs'.[43]

According to Sachs, Vaillant made no contribution to the theory of the sexuality of plants and 'can only have the credit of an eloquent description of what was then accepted', which was not much.[44] But if what was accepted was, in some sense, the literal truth of plant sexuality, and Vaillant's contribution is to have disseminated this truth more widely, it is striking that he rendered it intelligible by means of several extended analogies with aspects of animal, and more specifically, human sexuality that are often a poor fit. Vaillant's analogies are partly fantasmatic: for example, the suggestion that the flower bud bursts open because of the swelling of the male genital organ, that the male plant organs are tumescent with desire and that release of pollen is akin to sexual pleasure and orgasm. Vaillant's vocabulary in the description of ejaculation also often evokes the explosion of a gunshot or canon (the anthers/testicles being powder cartridges). The analogy between the plant and animal sexual organs often does not work (an animal's testicles do not explode to release semen) and the idea of 'amorous exercises' is stretched very far indeed.

Despite their limitations, these analogies are extended even further in Linnaeus' writings on the sexes of plants and in the terminology of his sexual system for the classification of plants. In the earliest of these writings, his *Praeludia Sponsaliorum Plantarum* (*Prelude to the Betrothal of Plants*), 1729, Linnaeus says that he had not actually read Vaillant's *Discourse*; in a later autobiographical sketch he says he had, however, read a review.[45] Whatever his source it was sufficiently detailed to allow Linnaeus to explain and then use Vaillant's vocabulary for the male and female parts of plants (ovary and '*tuba*' – Vaillant's '*trompe*' – for the female parts and *testiculos* and *vasa spermatika* for the male parts).[46] It is not clear whether Linnaeus knew of Camerarius's, *Letter* at this time (though he might have known via his apparently second-hand knowledge of Vaillant).[47] But having briefly mentioned various experiments proving that fertilization takes place '*per testiculos*' via the '*farina seminalem*', he concludes that 'no one appears able to deny the existence of sex in plants',[48] as if to affirm the literal interpretation of the sexes in the *Letter*.

However, Linnaeus's *Prelude* is structured around, and explicitly notes, a series of analogies with animal life (in the prefatory remarks he writes: 'These few pages deal with the great analogy that exists between plants and animals in terms of their similar means of propagating their species').[49] Having distanced himself from the mistaken view of the ancients,[50] Linnaeus's argument begins with the claim that botanists have noticed many more analogies between animals and plants and that Malpighi and Grew, in particular, brought many analogies to our attention.[51] The search for the generative organs of the plant is then an extension of this analogical method. Beginning with the premiss that nature is always consistent with itself, he writes that as plants bear fruit the male organs in the plant must be those that impart life to the fruit, just as in the animal kingdom the male sperm (*genitura maris*) is necessary to the production of an embryo.[52] This is followed by another basically analogical argument. We know, he says, that the flower always comes before the fruit, and that the fruit is 'its real offspring', so there is no fruit without the prior flower just as there is no embryo in the animal kingdom without 'previous union'. Thus the flower is a necessary condition for the fruit, as the male and female genital organs in the animal are the necessary condition for the embryo, and it follows (according to Linnaeus) that the organs of generation must be in the flower.[53] The male and female organs of the flower are then identified according to a further analogy with animals drawn from Vaillant. The ovary is identified as such because 'it performs the same service as the ovary in the animals', and the tuba 'from the analogy that exists with the Fallopian tube in the animal kingdom'. The *vasa spermatica* (stamens) and the *testiculos* (anthers) are called male because from them comes the pollen which is 'a substitute for seminal fluid [*succedaneum geniturae*] which fertilises the ova'.[54] This emphasis on the *organs* of generation is, as we shall see later, particularly important.

The basic parts and sexes having been established, Linnaeus continues the analogy in a more fanciful way whilst, at the same time, identifying the terms of the analogy in such a way that the distinction between the literal and the analogical collapses. In the explication of the figures showing the parts of the flower he identifies the stigma as the vulva and the pistil as the vagina. That is, the parts of the flower on the diagrams are not just compared analogically to vulva and vagina; they are given these as scientific names (see Figure 4.1).[55]

The description of the other parts of the flower and their relation to the sexual organs evokes the classical epithalamium, the poem written for the bride on the way to the bridal chamber (in the diagram of parts the petals are named, as if scientifically, 'Thalamus', 'nuptial chamber'[56]):

The actual petals of the Flower contribute nothing to generation. Instead, they serve only as Bridal beds that the great Creator has so gloriously provided, adorned with generous Bed curtains and perfumed with many delightful fragrances, where the bridegroom and his bride may celebrate their Nuptials with so much greater solemnity. Now that the bed has been thus prepared, it is time for the Bridegroom to embrace his beloved Bride, and offer her his gifts. By this I mean that one sees how the testicles open and pour forth the generative powder, which falls on the tube and fertilises the ovarium.[57]

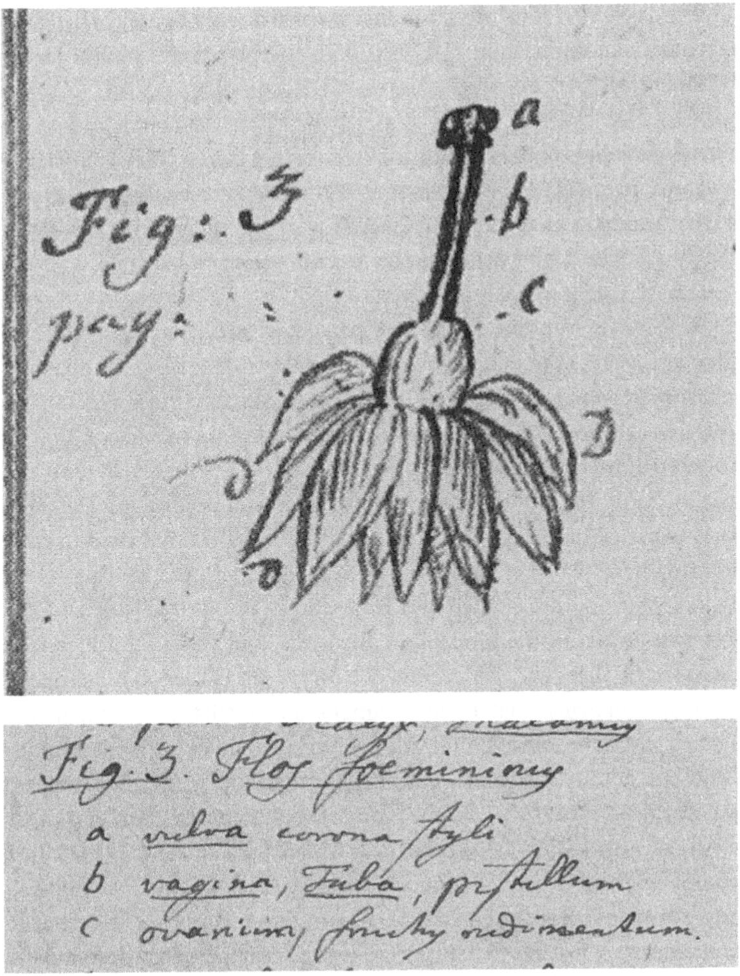

FIGURE 4.1 Linnaeus's drawing of the 'female flower' and identification of parts, in *Praeludia sponsaliorum plantarum* (1729). The original belongs to Uppsala University Library (D63).

Linnaeus's *Prelude* was written for his professor, Olof Celsius, in lieu, as Blunt tells us, of the verses usually presented on New Year's Day. Only twenty-two years old, Linnaeus wrote in the foreword:

I am no poet, but something, however, of a botanist; I therefore offer to you this fruit from the little crop that God granted me... In these few pages I treat of the great analogy which is to be found between plants and animals, in that they both increase their families in the same way.[58]

But the *Prelude* is not a youthful poetic work that can be filed as Linnaeus's juvenilia, in contrast to his serious scientific work. Its analogical method is characteristic of much of Linnaeus's work throughout his life,[59] and the poetic naming persists into the main divisions of his famous sexual system for the classification of plants.

In *Systema Naturae*, in the one-page 'key' to the sexual system in the first (1735) and all subsequent editions, the all-encompassing category is 'Marriages of Plants' (*Nuptiae Plantarum*)[60] (see Figure 4.2). The first division is between public and clandestine marriages: that is, between flowers visible to everyone and flowers 'scarce visible to the naked eye'.[61] The only class in the division 'clandestine marriages' is Cryptogamia, hidden or secret marriages – the word still used to refer informally to plants (e.g. ferns) that reproduce without bearing seeds. Public marriages are divided into Monoclinia ('In one bed') and Diclinia ('In two beds'). In the Monoclinia husbands and wives rejoice or take pleasure in the same bed chamber (*Mariti & Uxores uno codemque Thalamo gaudent*); that is, the stamens and pistils are found in the same (hermaphroditic) flower. Monoclinia is divided into those groups 'Without affinity' (*Diffinitas*), where the husbands (i.e. the stamens) are not related (*cognati*) or not joined together at any part, and those 'With affinity' (*Affinitas*), where the husbands are 'related'. The Diclinia comprises three classes: Monoecia ('One house'), Dioecia ('Two houses') and Polygamia (Many marriages). Again, these words are still used in contemporary botany. In Linnaeus's key the Monoecia are described as those plants in which male and female share the same house but live in different chambers (*Mares habitant cum fem. in eadem domo, sed diverso thalamo*) – that is, where separate 'male' and 'female' flowers are found on the same plant. In the Dioecia they live in different houses (male and female flowers on different plants). The Polygamia are those with male, female and hermaphroditic flowers in the same or different plants, or where 'Husbands live with wives and concubines' (*Mariti cum uxoribus & innuptis cohabitant in distinctis thal*). This polygamy can be 'Superfluous', when married females are fertile so the concubines or prostitutes (*Meretrices*) are superfluous; 'Frustraneous' (or 'Frustrated', *Frustranea*), when the married females are fertile and the concubines or

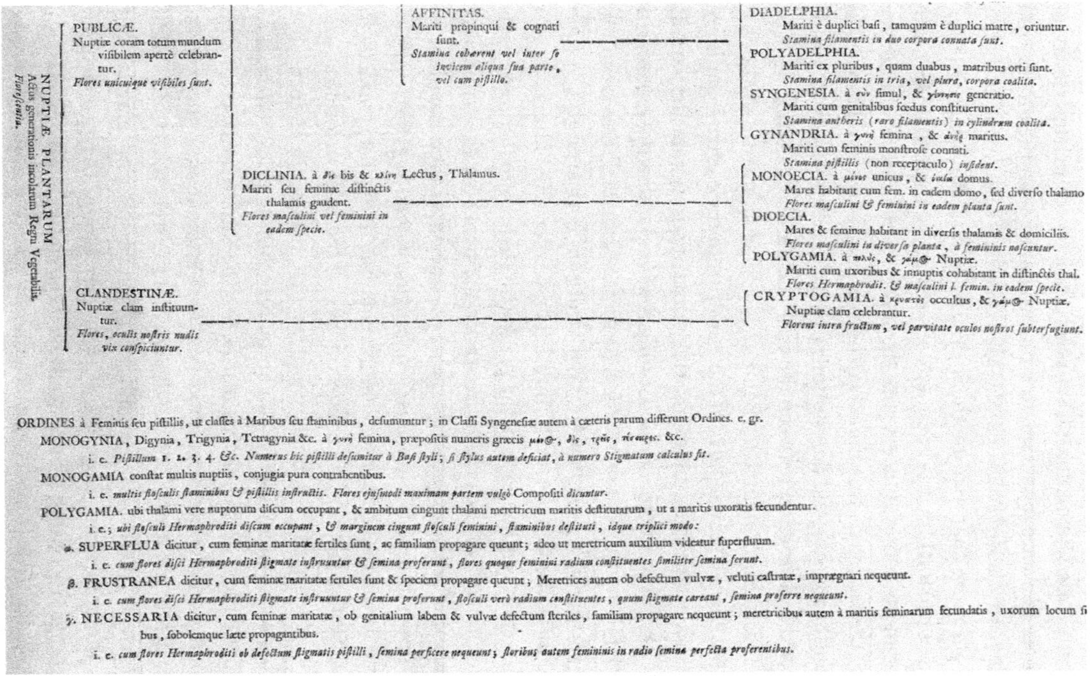

FIGURE 4.2 Detail from Linnaeus' s'Clavis Systematis Sexualis', the 'Key' to the sexual system, in *Systema naturae*. By permission of the Linnean Society of London.

prostitutes are barren; or Necessary (barren wife, fertile concubine). The twenty classes of the Monoclina are, as is well known, named according to the number of stamens and their relations to each other, for example, the simple first class of Monandria (One male), the more complex fourteenth class Didynamia ('Two powers', or power or superiority of two; where there are four stamens with two longer than the others) and the eighteenth class of Polyadelphia (Many brothers), where the husbands arise from more than two mothers (*Mariti ex pluribus, quam duabus, matribus orti sunt*); that is, 'Stamens are united by their filaments into three or more bodies'. Once again, then, the embedding of the source terms of the analogy – this time from human social-sexual relations – in the scientific naming and descriptions of the classes and orders of plants collapses the distinction between the analogical and the literal.

In 1736, in his *Fundamenta Botanica* – a short aphoristic textbook – the names for the sexual parts of the flower from the earlier *Prelude* are given as alternatives to the more widely used Latin names for various parts: 'Calyx ergo est *Thalamus*, Corolla *Auleum* [a hanging tapestry or curtain]; Filamenta *Vasa spermatica*, Antherae *Testiculi*, Pulvis *Genitura*, Stigma *Vulva*, Stylus *Vagina*, Germen *Ovarium*, Pericarpium *Ovarium foecundatum*, Semen *Ovum*'.[62] Immediately after this Linnaeus also identifies various other parts of the plant with their animal counterparts (e.g. the stomach of the plant is the earth, the bones are the trunk, and the lungs are the leaves) and notes that the ancients called the plant an upside-down animal.[63] Linnaeus's *Philosophia Botanica* (1751) follows the same format as the earlier *Fundamenta* but is much expanded. As Frans A. Stafleu notes, the chapter on 'Sex' is the least taxonomic and the most 'biological'.[64] Linnaeus affirms that plants are indeed alive through a series of analogies with animals including origin (from seed or egg), nutrition, age (such that plants, too, are said to have an infancy, childhood, adolescence, manhood [*virilitas*] and old age), movement, disease, death, anatomy (seemingly ascribing vessels, sacs, ducts, skin and cuticle to plants) and organism (plant vessels and glands).[65] It repeats the arguments from the *Prelude* concerning the origin of all living things, including plants, from an egg, and extends the analogy with animals.[66] He repeats the arguments showing that the flower contains the genital of the plants, further specifying that the ripening of the fruit is the birth or offspring (*Partus*).[67] He explains the need for pollination, mentioning the controversies over whether the pollen grains enter the ovary and coming down (though not decisively) on the side of the idea of 'the *seminal breeze*' (*auram seminalem*).[68] He repeats the analogies with the parts of animals (bones, lungs, heart and so on) and all parts of the 'bedroom' analogy from the earlier texts, with the further detail that the Calyx (already identified as 'Thalamus') 'could also be regarded as the *lips of the cunt* or the *foreskin* [Calyx posset pro *Cunni labiis*, vel *Praeputio* etiam haberi]' and the corolla with the '*nymahae*', that is, labia minora.[69]

The extended anthropomorphic analogy that structures Linnaeus's understanding of fructification and of the system of classification based on it also involves the familial terminology of vegetable genus-species relations – 'mother', 'father' and 'daughter'. This terminology is used in the 1760 *Dissertation on the Sexes of Plants*[70] and in Linnaeus's lectures of 1764 and 1771. In the lectures it is seemingly a response to the problem of the growing implausibility of maintaining his early views on the fixity of species. Accepting the need to introduce historical (i.e. genealogical) factors Linnaeus's lectures taught that 'all species of one genus have arisen from one mother through different fathers' and that these species 'should be referred to the mother's genus as her daughters'.[71] It is also related to the attempt to make the avowedly artificial system more 'natural' (avoiding the violation of natural groupings through the too-strict adherence to a method of classification through a single character).[72] Linnaeus thus aims to make his theory more scientifically adequate by characterizing the relations between genera and species with the terms of human kinship. In all cases, then, the source terms of the analogy between animal/human and plant *become* the scientific designation of various categories and genealogical relations, such that the distinction between the analogical and the literal becomes very difficult to sustain.

The ambiguous 'male'

Bearing Vaillant and Linnaeus in mind how, then, do we understand the claim that progress is made in the theory of the sexuality of plants when the discourse shifts from analogy to literal identification? Both thinkers explicitly maintain a series of analogies with animals and identify a series of correspondences between parts and functions on that basis, so the analogical *method* is certainly not rejected. In contemporaneous texts and in the reception of Linnaeus's work, especially, other authors also continued to present plant sexuality (and many other aspects of vegetable life) principally via an analogy with animals.[73] When, in these texts, various authors are credited with having discovered 'a True Notion of Sexes of Plants', or the 'true sexual Distinction', this is a contrast with the *metaphorical* masculine and feminine characteristics according to which the ancients called some plants (mainly trees) male and female, as noted by Theophrastus.[74] It is not a contrast with an analogical approach to plant sex; what is 'true' is the location of the analogy in the parts or organs of the flower or in the sexed individuals.

Perhaps an historian like Sachs would say that, in these texts, the analogical method allows for the discovery that plants are literally sexed male and female, to the extent that the necessity for both in the generation of fertile seed is affirmed, but that this literal, scientific truth is accompanied (adorned or disfigured, according to one's view) by a purely rhetorical *metaphorical* apparatus. The literal truth could then be extracted for science, while the metaphorical aspect could be seen as being developed in the more obviously philosophical and literary reception of Linnaeus, notably Julien Offray de la Mettrie's *L'homme plante* (*Man as a Plant*) (1748) and Erasmus Darwin's well-known poem *The Loves of The Plants* (1789). La Mettrie imagines 'man… metamorphosed into a plant',[75] where 'man' stands in for the animal kingdom as a whole. In it one finds all the usual analogies but more besides. For example:

> One can regard the virgin womb, or rather the womb not pregnant, or, if you like, the ovary, as a seed which is not yet fecundated. The *stylus* of the woman is the vagina; the vulva and the *mons veneris*, with the odours which the glands of these parts give off, corresponds to the *Stigma*: and these things, the womb, the vagina and the vulva form the *Pistil*; the name that modern botanists give to all of the female parts of the plant.[76]

La Mettrie compares human men to Linnaeus's Monandria: 'For us men, one glance is enough: sons of Priapus, spermatic animals, our *stamen* is as if rolled into a cylindrical tube, that is the *rod* [*verge*; penis], & the sperm is our fecundating *powder*.' Women are Monogynia 'because they only have one vagina'.[77] In Erasmus Darwin's *The Loves of The Plants* the analogy of parts as well as the personification of parts as husbands and wives is only the starting point for a long poem – presented as the narration of the 'Botanic Muse' who 'in this latter age/Led by [her] airy hand the Swedish sage'.[78] Representatives of the Orders of Linnaeus's system are cast in vignettes of amorous behaviours with recognizable types (blushing maid, wanton harlot, siren and so on) and their beaux or victims. For example:

Sweet blooms GENISTA in the myrtle shade,
And *ten* fond brothers woo the haughty maid,
Two knights before thy fragrant altar bend,
Adored MELISSA! And *two* squires attend.[79]

La Mettrie and Darwin could be contrasted with the scientific popularizers of Linnaeus's system, in England especially, who, as Londa Schiebinger says, made little use of the sexual imagery[80] – or who, at least, did not elaborate it in the manner of La Mettrie or Darwin. William Withering, in his *Botanical Arrangement of British Plants* (1776), used Linnaeus's system but, he writes, 'it was thought proper to drop the sexual distinctions in the title to the Classes and Orders', so that, for example the description of Monandria Monogynia becomes '*Thread single… Shaft single*'.[81] Richard Pulteney, in his *A General View of the Writings of Linnaeus*, was similarly uneasy:

Linnaeus was the first who constituted the *stamina* and *pistilla* the bases of an *artificial* method of arranging plants; and he tells us, in his CLASSES PLANTARUM, that he was led to it by considering the great importance of these parts in vegetation. They alone are essentially necessary to fructification, all other parts, except the *antherae* and *stigmata*, being wanting in some flowers; and the present philosophy of botany regards the former as the *male*, and the later as the *female* organs of generation. As such indeed they must be considered, *analogically*, yet perhaps the Linnean system, admirable as it is, would not have been less acceptable, had the classical terms been expressive only of *number* and *situation*, without regard to the offices of the parts, in framing the terms.[82]

Priscilla Wakefield's *An Introduction to Botany* (1796) also described the fructification (flower and fruits) using the older English words (e.g. 'chives' and 'style' or 'pointel') and followed Linnaeus's system but without sexing the parts.[83]

These texts, then, seem to provide the basis for an interpretation of eighteenth-century developments as the progressive extraction of the literal core of truth from the metaphorical presentation of Linnaeus, especially. But there are fundamental problems with the attempt to distinguish between the literal truth of plant sex known to modern science and its metaphorical expression in Vaillant and Linnaeus, or between a literal and a metaphorical understanding of it.

First, quite generally, the absolute distinction between the literal and the metaphorical has not survived either its analysis in the history of the sciences or its deconstruction in philosophy.[84] Second, the texts themselves are far from affirming a distinction between the literal and the metaphorical. Indeed, what is most characteristic of the texts themselves is a profound and arguably ineliminable ambiguity between what is literally and metaphorically meant, especially in the incorporation of the 'metaphorical' terms into their scientific vocabulary. The success and persuasive force of the texts lies precisely in this. It might be objected that the extension of the vocabulary of marriage and so on into the discussion of plant reproduction is an example of what Max Black calls 'catachresis': 'the use of a word in some new sense in order to remedy a gap in vocabulary. Catachresis is the putting of new senses into old words'. This catachrestic use of metaphor 'plugs the gaps in the literal vocabulary' of a science, and 'if a catachresis serves a genuine need, the new sense introduced will quickly become part of the *literal* sense... It is the fate of catachresis to disappear when successful'.[85] Richard Rorty similarly argues that to understand a metaphor, to grasp the meaning conveyed by it, is not to understand some special kind of metaphorical meaning but 'to de-metaphorize the sentence and endow it with a use... a new *literal* sense'.[86] According to this view, as Mary Hesse points out, a dead metaphor is really 'a new stage of literal language'.[87] But while it may indeed be true that the language is taken to be and used *as if* literal, there is still no ground for claiming a pure literality of language that is distinguished and distanced absolutely from metaphor.

The same constitutive ambiguity between the literal and the metaphorical is still characteristic of much of the modern scientific vocabulary used in the discussion of plant sex. As we indicated in our discussion of Camerarius, from the standpoint of the modern definition of sexuality, what Vaillant and Linnaeus think of as 'sexes' are so only in a manner of speaking, *not* literally. Any dioecious plant itself, any flower on any plant, any stamen or pistil in any flower can be *called* 'male' or 'female', but they are not literally such, according to the modern definitions. The extensive and almost unquestioned use of the terms 'male' and 'female' in this way in modern botany and horticulture thus still employs the basic analogy with animals to the extent that it postulates as functional counterparts the sexual organs of animals and the parts of a flower or equates sexed animal individuals with sexed flower or plant individuals. When we do this today we draw an analogy that does not, in fact, point out a literal correspondence, because the stamen is not a penis, the anther is not a testicle, and nothing in a plant corresponds to an animal vulva.

The attempt to distinguish between the literal truth of plant sex and its metaphorical expression in Vaillant and Linnaeus is further complicated by the scientific vocabulary that Vaillant and Linnaeus proposed. This is particularly obvious with Linnaeus, both when the analogical naming of parts transforms frictionlessly into their scientific designation (the parts of the flower are *labelled* 'vagina' and 'vulva') and in the embedding of the metaphorical aspects of his personification of plant amours into the very names of the plant groups. Some of this vocabulary is still in use in modern botany: in the terminology used to name (non-phyletic) groups of plants (e.g. polygamous, monoecious, dioecious); in the terminology drawn from the idea of 'marriage' (e.g. cryptogams, dichogamy, gametes, syngamy) and in the familial terminology of mothers, fathers, daughters, sisters and so on. Perhaps it will be said that these are 'dead metaphors'? But, as Evelyn Fox Keller writes, 'metaphors are dead only because we cease to notice them, because we are no longer conscious of their effects on our perception. It might even be argued that dead metaphors are the most forceful of all, just because their operation is beyond the realm of consciousness, effectively screened by their very banality'.[88] More undead than dead, then, these zombie metaphors.

In both these respects the compounding of the literal and the metaphorical (or of similarity and identity) is in part an effect of the equivocation between function, organ and individual in the reference to 'male' and 'female'. We can illustrate this with a contrasting example. Let us imagine an analogy between the lungs of an animal and the leaves of a plant. In the case of this analogy, we would distinguish terminologically between the *function* of breathing or respiration, the *organ* that performed the function (lungs or leaves), and the *individual* organism that was possessed of the organ (the lunged animal or leaved plant), whom we might call the breathing individual. Having so distinguished the elements the organ would not be conflated with the breathing individual; we would not call the individual organism the lung in anything except a clearly metaphorical sense, as when trees are said to be the lungs of planet earth. In the case of sex, the elements can be designated differentially to refer (using the example of the male) to the *function* of reproduction, the penis/testicles or stamens/anthers as the *organ* of reproduction and the *individual* organism (animal or plant) to whom the organs belong. But precisely because of the difference of sex, each element requires and can be referred to in terms of its adjectival qualification: *male* function of reproduction, *male* organ and *male* organism. It is the necessity for the adjective that then allows for the easy conflation of the elements: things that are not in fact the same (organ and individual) are both able to be referred to as 'the male'. (In fact, Linnaeus tried to avoid this in his *Philosophia Botanica*, distinguishing terminologically between the masculine or feminine (*masculus* or *femineus*) flower and the male or female (*mas* or *femina*) plant.[89] But the designation of the pistil as the vagina, for example, rather negates the effort.)

The addition of the familial terminology of human kinship relations tightens the conflation of organ and individual, as when, for example, Camerarius says that the female genital organ of the plant performs the maternal office (which *it* can only do if *she* is a mother). Vaillant and Linnaeus take the conflation of organ and individual (effectively, the personification of the organ) very far, and we would not expect to see that in biology today. But the same kind of equivocation between or conflation of elements does characterize reference to and even definitions of sex categories in modern biology where 'the male' is both the gamete and the organism that produces the gamete, that is, the 'individual' organism, or where the gamete becomes the effective 'individual' when its progress and behaviour becomes the topic of research and discussion. As we saw in Chapter 2, Aristotle was careful to distinguish between the different meanings of 'male' and 'female' (metaphysical principles of generation, individuals and organs or parts). For him, in any discussion of whether organisms have male or female it was imperative to first specify in what sense one understood those terms. Although we know that that same specification is presupposed in the scientific literature and that the literature devoted to the process of sexuality, specifically, is required to proffer its definitions explicitly, in general biological discourse (and perhaps especially in botany) the conflation of the different senses of 'male' and 'female' is commonplace, as it is in everyday speech.[90]

Once again, we see how it is in or through the study of vegetable life that the question of the meaning of 'male' and 'female' across different forms of life is brought to our attention. In Vaillant's and Linnaeus's discussions of plant sex the constitutive ambiguity of the literal/ metaphorical is the mark of the conceptual vacillation of 'male' and 'female' between function, organ and individual, the last two especially. Their texts play on the different meanings of sex for rhetorical purposes, but also allow us to see how the field of discourse on sex and sexuality is always, and perhaps ineliminably, scored across by these different meanings. This is particularly obvious in discussions of plant sex because there the problem of the meaning of 'male' and 'female' becomes explicit. But (as we shall see further in the following chapters) it is no less true of discussions of sex in general, including in modern botany, because of ambiguities arising from the shared terminology of sex for functions, gametes, organs and individuals. Modern (re)definitions of 'male' and 'female' attempt to overcome these ambiguities, but in the case of sex it is particularly difficult to do away with ancestral meanings, and particularly difficult to separate scientific definitions of sex and of male and female from the popular meaning of these terms. The following two chapters and our Epilogue explore this difficulty further and attempt to propose a philosophical account of the constitutive ambiguity of 'sex' that, while not doing away with the problem, does at least understand it in greater depth.

5 WHAT ARE 'MALE' AND 'FEMALE' IN PLANTS?

We can summarize the arguments of the previous three chapters, very briefly, as follows. Traditional histories of botany, written from the standpoint of the present, are surprised that it took philosophers, naturalists and scientists so long to discover the 'obvious' truth of plant sex. But from our plant philosophy perspective the hesitations and doubts of Aristotle and the thinkers of the Aristotelian botanical tradition are not historical confusions or errors; rather, they are the consequence of a properly philosophical approach to the issue. The idea of plant sex was not obvious for these thinkers because an answer to the question 'Do plants have "male" and "female"?' required an answer to another, more fundamental philosophical question: 'What are "male" and "female"?' A first response to this more fundamental question can be phrased in a characteristically Aristotelian form: 'male' and 'female' are said in many different ways. Aristotle's distinctions between the metaphysical principles of male and female, male and female individuals and male and female parts or organs allowed these thinkers to discuss the ways in which plants could be said to have male and female: it was not a question with a 'yes' or 'no' answer. Part of this involved an explicit and critical reflection on the analogy with animals, not taking it for granted.

Whereas, then, the traditional histories of botany see the philosophical aspects of that history as a regrettable drag on scientific progress, from our plant philosophy perspective this is part of its strength. As we showed in Chapter 3, sixteenth- and seventeenth-century botany (with Cesalpino and Grew in particular) progressed partly *through*, not despite, its philosophical commitments. Concomitantly, philosophy proceeds through botany, in discussions of plant sex, because of the necessity, there, to reflect on the different meanings of male and female. The traditional histories of botany also celebrate the entry into the properly scientific era of plant science with the acceptance of the literal truth of plant sex in the late seventeenth and eighteenth centuries, a progression which they see as dependent upon the jettisoning of the metaphysical baggage of the earlier tradition. But, as we have seen in Chapter 4, Vaillant's and Linnaeus's discussions of plant sex are in fact notable for the literal/ metaphorical ambiguity of their terms, both in their descriptions and in their scientific terminology, an ambiguity which conditions and is conditioned by a collapse of the Aristotelian distinctions between the different meanings of sex, or a conflation of the different meanings. This collapse is less a jettisoning of metaphysical baggage than a philosophical regression. Because of the allegedly literal meaning now ascribed to male and female in plants, the limits and effects of the founding zoological analogy are no longer an explicit concern and the question of the being of 'male' and 'female' recedes. So, in one sense it is true that philosophy was (on the surface, at least) banished from botanical science. But what did botany thereby lose? What did philosophy lose when it apparently accepted that plants were no longer part of its intellectual domain? What if both botany and philosophy still need to ask, together: 'What do we mean by "male" and "female" in plants?'

As we mentioned in Chapter 1, the dibiontic life cycle (or 'alternation of generations'), unique to plants, complicates the identification of male and female in these forms of life. In this chapter we will look in a little more detail at the specificity of the dibiontic life cycle, which requires a brief scientific presentation. We will see how the recognition of the alternation of generations in plants once again explicitly raised the problem of the identification of male and female in plants, a problem which played out at the end of the nineteenth century and the beginning of the twentieth in a terminological debate in botany over the use of 'male' and 'female'. With this debate in mind, we will then look at some of the different usages of sexed terminology in more recent scientific and popular scientific literature on plant form and plant sexual reproduction. We will see that in this literature deviation from the strict scientific usage is surprisingly common; 'male' and 'female' are, in fact, used in several different senses, sometimes simultaneously. In some areas of plant science, indeed, a not-strictly-scientific use of the terms 'male' and 'female' is axiomatic, both textually and conceptually, as we will see. In certain respects, then, discussions of plants maintain a doubled discourse on male and female, overlaying a common-sense zoological model onto the strictly biological plant model. What are the implications of this?

In this chapter we will argue that attention to the dibiontic life cycle of plants, far from revealing anything obvious about plant sex, rather allows the full complexity of the problem to emerge. Attempts to resolve the terminological dispute about 'male' and 'female' in plants, and to propose new terminologies or definitions to describe the different sexual forms and sexual systems of plants, are, of course, based on the need to acknowledge biological reality. But they are equally – if not more so – either pragmatic compromises or attempts to construct new concepts of sex (and gender) and male and female for plants. It is this construction that is of particular interest to plant philosophy. To demonstrate this, we will focus particularly on the influential work of the botanist David G. Lloyd, from the 1970s and early 1980s. The kinds of conceptual and terminological problems and questions that Lloyd's work addresses already arise, as we will show, in Charles Darwin's *The Different Forms of Flowers on Plants of the Same Species* (1877), a text to which Lloyd often refers. Lloyd's proposed botanical definitions of 'sex', 'gender' and 'male' and 'female' also arise out of an acknowledgement (although it is perhaps not wholly explicit) of the inadequacy of the animal model for the description and understanding of plant populations and their sexual reproduction. In this chapter we will argue that this kind of definitional work has an ineliminably philosophical aspect, and that Lloyd's work on this topic, as well as being a contribution to plant science, can also be interpreted as contributing to a philosophy of plant sex.

Dibiontic life, or alternation of generations

First, though, it is necessary to appreciate the specificity of the dibiontic life cycle of plants, which is also to appreciate the specificity of the forms of sexual reproduction in plants. The phenomenon of the 'alternation of generations'[1] in plant life cycles was first identified in 1851 by William Hofmeister in his studies of the 'higher cryptogamia' (mosses and ferns), drawing on his contemporaneous work on embryology in flowering plants. In 1872 Sachs described the results of Hofmeister's investigations in plant embryology and the higher cryptogamia as 'magnificent beyond all that has been achieved before or since in the domain of descriptive botany'. Hofmeister showed, in both flowering plants and cryptogams, that the egg cell is formed in the embryo sac prior to fertilization and that its development into an embryo is dependent on its relation to the pollen tube. This meant, according to Sachs, that 'the idea of what is meant by the development of a plant was suddenly and completely changed' with alternation of generation 'proved to be the highest law of development', demonstrating a developmental commonality in otherwise very different plants such as liverworts, mosses, ferns and angiosperms (flowering plants).[2] What Sachs and others after him admired was not just the intellectual achievement of the discovery of the alternation of generations but also to some extent the unexpected and marvellous phenomenon itself. They admired, as Nils Svedelius put it, 'that singular relation of alternation in the ontogenetic development of plants which [Carl Nägeli] called "one of the most remarkable phenomena of the plant world"'.[3]

The alternation of two distinct, multicellular generations in plants is now understood primarily in terms of the different ploidy of those generations, that is, the different number of sets of chromosomes. (Even the simplest presentation of this cannot avoid the fact that it is rather complex, so we must impose a couple of pages of technical vocabulary on the reader here.[4]) The haploid generation, whose cells have one copy of each chromosome, is often abbreviated as $1N$; the diploid (with cells with two sets of chromosomes) as $2N$.[5] Before looking at this, it may be useful to be reminded of the typical animal life cycle, in contrast with which the specificity of alternation of generations in plants becomes clearer.

Animals – humans, for example – are multicellular diploid organisms; that is, their cells contain two sets of chromosomes, one set from each parent. In growth and maintenance (cell replacement) these cells divide mitotically, mitosis being a process of cell duplication (one cell divides into two genetically identical 'daughter' cells). The only haploid cells in the animal organism are the gametes, the ova or the spermatozoa, sometimes called the sex cells or reproductive cells. Gametes are produced (in the diploid organism) by the extraordinarily complex process of meiosis, the nuclear division (halving the number of chromosomes) that occurs after the 'pairing' of the maternal and paternal chromosomes in the cell and the 'crossing over' (swapping of DNA sequences) at specific points, or genetic recombination. The fusion of two haploid gametes (syngamy) in sexual reproduction produces a zygote, in which the diploid state is restored. The diploid zygote becomes the diploid embryo which (all being well) is born, grows, lives and dies as a diploid organism, itself usually able to produce haploid gametes. Animal gametes, some of which are motile and ejected from the body, cannot live independently of their diploid producer and do not grow into a haploid organism. The animal life cycle is therefore said to be 'monobiontic' – there is only one multicellular diploid generation.

In the dibiontic life cycle of the plant, on the other hand, there are two multicellular generations, which may be more or less physically independent of each other. One generation is diploid, the other haploid.[6] Whereas in the animal life cycle the diploid organism produces the haploid gamete by meiosis directly, in the dibiontic life cycle the diploid generation produces a haploid generation that produces the haploid gamete. The diploid generation is most often (mosses, hornworts and liverworts are exceptions) the plants that we would recognize. They are, as L. H. Bailey put it, what we would think of as 'the plant'[7] itself – the tree on the street, the grass in the meadow, the fern uncurling in my garden, the flowering bush, the dandelion pushing up between the cracks in the pavement. This diploid generation (or phase) is the 'sporophyte' (literally the 'spore-plant'), which produces, by meiosis, spores. That ferns, for example, produce 'spores' is well known, not least because the spore cases (sori) are visible on the underside of leaves. It is far less obvious, because not easily seen, that the sporophyte generation of flowering plants also produce, by meiosis, microspores (in the anthers) and megaspores (in the ovule). These haploid spores divide mitotically to produce a second (haploid) generation – the 'gametophyte' (literally the gamete-plant) which produces (haploid) gametes by mitosis.

What is the difference between a spore and a gamete? Gametes can fuse with other gametes; indeed, with very few exceptions, if they do not undergo syngamy they die.[8] Spores cannot fuse with other spores but they can divide mitotically and grow into a new organism – the haploid plant or gametophyte.[9] In most cases the sporophyte and the gametophyte do not resemble each other at all; the generations are heteromorphic. Before the discovery of chromosomes, the discussion of the alternation of generations was thus at first the description of the alternate *morphologies* of the two generations, which is easy to see in ferns and mosses with the naked eye. (Ironically it is the flowering plants which are the 'cryptogams' in this respect.) Figure 5.1 shows, in a schematic form, the life cycle of the fern (or an idealized version of it). The sporophyte generation is the 'the fern' that most people would recognize; the gametophyte is represented by the smaller, heart-shaped plant. Although the gametophyte is often

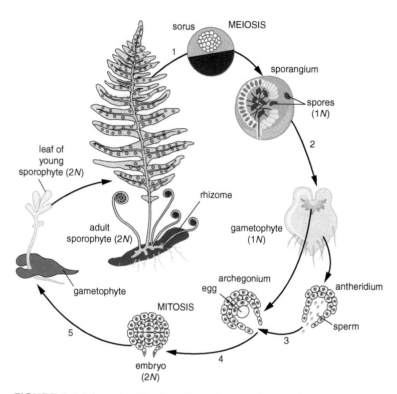

FIGURE 5.1 Life cycle of the fern. Reproduced by Integra Software Services.

very tiny, a mature gametophyte is an 'adult' plant. The male gametes are produced in a region of the gametophyte called the antheridium, the female in the archegonium. Ejected sperm (as they are called) swim through surface water and enter the archegonium where the eggs are produced. If fertilization occurs (gamete fusion produces a zygote that becomes an embryo) a new diploid sporophyte will grow and the haploid gametophyte dies. With mosses, on the other hand, the 'mossy' green plant with which we are most familiar is the haploid gametophyte phase, spreading vegetatively. The sporophyte generation is often visible as the thin hair-like stalks that grow out of and remain attached to the mossy gametophyte until they die off.

The same alternation of generations characterizes the life cycle of flowering, seed-producing plants (angiosperms) (Figure 5.2). The diploid sporophyte is the recognizable flowering plant itself. Its haploid

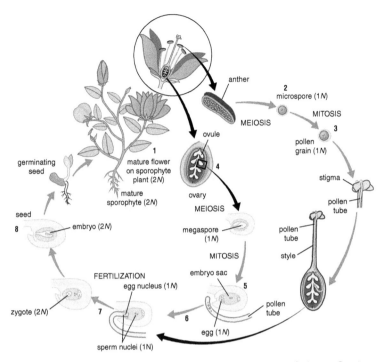

FIGURE 5.2 Angiosperm life cycle. Reproduced by Integra Software Services.

spores, the product of meiosis, are the pollen grains produced in the anthers (the microspores) and the megaspore produced in the ovule (the ovule being a part of the plant, which is itself part of the ovary; we will discuss this terminology in Chapter 6). The haploid microspores (pollen grains) develop into haploid microgametophytes comprising only two or three cells – so small that they fit inside the pollen grains. The haploid megaspore develops within the ovule into a seven-celled haploid megagametophyte for which the technical term is the 'embryo sac'.[10] The gametophytes then produce gametes by mitosis. Although it may seem counterintuitive, the microgametophyte and the megagametophyte are, like the independent gametophytes of ferns, entire and mature plants – a distinct, multicellular generation from the sporophytes which produced them. The haploid angiosperm gametophyte is 'a tiny mass of cells with no roots, stems, leaves or vascular tissues, but it is an entire plant'. And this is the case even though the megagametophyte develops *within* the sporophyte plant – it is 'one plant growing inside another'.[11]

The microgametophyte contains two sperm cells or microgametes. If a viable pollen grain from an anther arrives on a stigma it may germinate to produce this gametophyte generation within itself (within the pollen grain) and produce a pollen tube that grows rapidly down through the style, carrying the sperm cells to the ovule and then to the megaspore, which develops into the megagametophyte. One sperm cell (microgamete) and the egg cell (or 'ovum'; the megagamete) undergo syngamy to produce the diploid zygote that becomes the embryo and will eventually grow into the next generation (diploid) sporophyte. The second sperm cell fuses with the binucleate 'central cell' of the megagametophyte to produce the triploid endosperm. Because two sperm cells undergo fusion with different cells from the megagametophyte, the process is known as 'double fertilization'.[12] The life cycle of 'the' flowering plant and 'the' fern or moss thus involves at least three or four distinct multicellular organisms – minimally one sporophyte and two gametophytes[13] – but the sporophyte (a tree, for example) may, via the vast number of spores from its many hundreds of flowers, give rise to many hundreds of gametophytes in each flowering season.

These are the basics of plant sexual reproduction, the details of which can be found in any textbook of botany and in multiple iterations online.[14] What interest might plant philosophy have in this?

'On the untechnical terminology of the sex-relation in plants'

The fact of the alternation of generations in plants raises, once again, the question of what we mean by 'male' and 'female' in plants, and once again refers us to the prior question: 'What are "male" and "female" more generally?' Almost from the moment of its first description, the alternation of generations in plants was characterized as an alternation of asexual and sexual generations. The sporophyte generation was (and sometimes still is) called the asexual generation because its spores (unlike seeds) are produced asexually (i.e. without conjugation of different cells) and because they give rise to a new plant without themselves being fertilized. The spore itself was thus also called asexual.[15] The gametophyte (as it is now known) is sometimes called the sexual generation because it produces the gametes through which plants reproduce sexually.[16] This means that the sexual generation arises from an asexual generation. It also means that, strictly speaking, neither the sporophyte (what we mostly recognize as 'the plant' or 'the tree') nor any of its structures – including the structures of the flower(s) – are 'male' or 'female'.

In 1896, after the discovery of chromosomes and the introduction of the terminology of 'gametes', the influential American horticulturalist L. H. Bailey reflected on the implications of the sporophyte generation being understood as asexual. He noted that the discovery of the alternation of generations and the new conception of the 'sex-relation' in plants had given rise to new terminology and to attempts to 'restrict or to specialize the use of such age-long words as male and female, sex and the like, when applying them to plants'. Bailey reminds the reader that the 'original conceptions of sexuality in plants… were borrowed and adapted very largely from analogy with the animal kingdom'. This analogy, he points out, gave rise to more than just the idea of male and female in general: 'The stamens were considered to be male organs of sex and the pistils to be female organs, the idea of the necessity of a conformed sex-member being evidently borrowed from a knowledge of animal morphology.'[17] Hofmeister's discovery of a 'sex-bearing' generation followed by a 'sexless' generation has thus raised the questions, Bailey says, of 'how far can we use the terms "male" and "female" and what must be the common language of the sex-relation in plants?' As the 'male sex phase' is essentially only the 'short span and function of the generative cell developing from the pollen grain' and the female the development of the gametophyte in the ovule, some now object to calling the stamen 'male' and the pistil 'female'. Bailey sees that this objection is scientifically and logically warranted. Indeed, he says that if we carry this scientific logic to its end, we would not call any part of the gametophyte except the egg cell and the sperm cell 'female' or 'male': 'it follows that we cannot apply the terms "male", "female", "sex", and the like, to plants, save in the very brief period during which impregnation is taking place.'[18] And what, he says, if this 'hypercriticism' were applied to animals too? There would be 'pandemonium', as 'One could not speak of the members of generation as sex organs, not of any animal as male or female'.[19]

Bailey's overwrought fears of pandemonium are interesting, and no doubt betray something of the broader ideological context in 1896. But when botanists (and especially those concerned with botanical education) did rather straightforwardly seek to reflect the existence of the asexual generation in their terminology the sky did not fall in. The American botanist and educator John Merle Coulter, in various works, made a concerted effort to correct (as he saw it) the misnaming of the parts of flowers as 'male' and 'female' and it is notable that he did this in textbooks and shorter books intended for students, including secondary (or high school) students. In his *The Evolution of Sex in Plants* (1914) he defines alternation of generations as the separation of a spore-producing and a gamete-producing generation:

> so that there are two kinds of individual in a life cycle.... The individual with the sex organs (gametophyte) may be called the sexual individual, and the one without sex organs (sporophyte) may be called the sexless individual. Alternation of generations, therefore, is the alternation of a sexual individual with a sexless individual in a life cycle. This means that it takes two individuals to complete a life cycle.[20]

The evolution of gamete production is at the same time, for Coulter, the evolution of sex, in the sense of the 'act of fusion' and the evolution of what will be identified as 'the sexes'.[21]

His discussion of this includes an implicit reflection on the meaning of 'male' and 'female' in general. The first (evolutionary) stage of gamete production, he notes, is called 'isogamy', a term which is justified to the extent that it refers to the morphological likeness of the two gametes, but there must also be a physiological differentiation. Isogamy 'shades into' heterogamy which 'must not be regarded as the gradual differentiation of *maleness* and *femaleness*, but the gradual appearance of differences that make the two sexes recognisable'.[22] 'At this stage in our knowledge', he writes, 'it is useless to ask what "maleness" and "femaleness" are, but it seems evident that they represent essentially conditions of the nucleus'. That is, we *call* heterogametes 'male' and 'female', but these are 'words rather than explanations', standing in for 'a definite physiological difference'.[23]

Having located the distinction of sex at the level of the (sex) cell, Coulter also describes the evolution of the sex organs of plants (the 'gametangia') as having moved through three stages: unprotected gamete-producing cells associated with water habitat (in algae) jacketed organs (antheridia and archegonia) associated with the land habitat (e.g. ferns) and the 'final elimination of antheridia and archegonia, associated with the complete dependence and reduction of the male and female plants [i.e. of the gametophytes]'. In flowering plants, the archegonia (that part of the fern gametophyte that produces the female gametes) is 'reduced to its essential sexual structure, the egg'.[24] It is very specifically the sexual *organs* of both gymnosperms and angiosperms that Coulter describes as having been 'completely reduced' (or indeed, in the case of the female organ, 'eliminated').[25] If we can speak of the sexual organs – that is, the gamete-producing cells – of flowering plants, for example, we must be referring to a part of the gametophyte or to the gametophyte itself. As the gametophytes of flowering plants have 'disappeared from ordinary observation' in becoming very tiny and encased in the spore or the sporophyte, 'the sex organs can only be discovered by the use of laboratory technique'.[26] Coulter writes that it is therefore no surprise that earlier botanists, who worked out that the stamens and pistils were both necessary in the life cycle of the plant, *misidentified them* as the sexual organs of the plants. However, 'to regard a flower as a sex structure and its stamens and pistils as sex organs is to misapprehend the situation. They belong to the sporophyte, which does not produce sex organs... To speak of male flowers and female flowers, as is so often done, is natural, but it is untrue'.[27]

Coulter insists on calling the structures that produce spores 'sporophylls' (a term that today is sometimes defined as referring only to the spore-bearing leaves of ferns and mosses): 'Before the real nature of these structures was known, the microsporophyll was called a *stamen*; the region of the microsporophyll bearing sporangia was called an *anther*; the microsporangia, *pollen sacs*; and the microspores, *pollen grains*'.[28] In his *Plant Studies* (1901), a textbook for secondary school students, he had already noted that 'stamen', 'pollen-sacs', 'pollen-grain' and 'pollen' were the names given to the structures of seed-plants before their function was understood, although there he also says that they are 'names... still very convenient to use in connection with the Spermatophytes' so long as we remember that they are really microsporophylls.[29] However, there is no justification for referring to cones or to stamens and pistils as 'male' or 'female' organs, as far as Coulter is concerned, not even for convenience – it is simply biologically wrong.[30] As an educationalist, Coulter objected to the false teaching involved in referring to the 'sexual organs' of flowers: 'It should be borne distinctly in mind that the stamen is not a sex organ, for the literature of botany is full of this old assumption, and the beginner is in danger of becoming confused and of forgetting that pollen grains are asexual spores'.[31]

Bailey, by contrast, claims that the fact that the pollen grain and its gametophyte ultimately come from the stamen means that the botanist continues to be warranted in calling the stamen 'male', and recommends the retention of the 'old-time attributes' of maleness and femaleness for the parts of the flower:

> 'Male' and 'female' never did and never can be made to express strict morphological homologies. An organ of an animal or a plant is male if it exercises the functions of paternity and not of maternity. The stamen is such an organ. Its entire office is that of maleness... The common language of sex has always dealt in analogies. There are perfectly good and sufficient technical terms to designate the homologies and the ultimate physiological processes.[32]

While acknowledging the strict scientific correctness of the restriction of the uses of the terms 'male' and 'female' to gametes, Bailey thus argues that we should nevertheless retain the 'untechnical' terminology, borrowed from zoology and analogically understood, for the parts of flowers, individual flowers and individual plants. In 1905 Charles J. Chamberlain made much the same argument, again explicitly acknowledging the technical correctness of restricting 'male' and 'female' to gametophytes but recommending that the botanist should follow the zoologists' practice and to continue to speak analogically of male and female sexual organs in the flower and male and female individuals.[33]

And so it has come to pass in botany, to this day. In the long century since Bailey's article the botanical consensus is that the sporophyte and the parts of the flower are not, strictly speaking, 'male' or 'female' because the more general biological definition of male and female has come to be restricted to the designation of anisogamy (having two gametes of unequal size) and to the organisms that produce such anisogametes.[34] Only the gametophyte qualifies. Nevertheless, in almost all scientific work on plant reproduction, and in its popular communication, the zoological analogy continues to hold sway and sporophytes and the parts of flowers are routinely designated 'male' and 'female' (and, of course, hermaphroditic). We should have to say, then, that the much-vaunted progress of botany from an analogical to a literal understanding of male and female in plants, as discussed in Chapter 4, is not what it seems, and the dominant way of speaking about plant sex is only misleadingly characterized as 'literally true'.

There are examples of strict attempts to stay faithful to technical biological definitions of male and female in discussions of plants. James Mauseth's widely used botany textbook explains that the parts of the flowers are not, strictly speaking, male and female and sticks thereafter to a technical – that is, specifically botanical – vocabulary that avoids the explicit zoological analogy. Monoeicy and dioecy are explained without reference to male and female, flowers are called 'staminate', 'carpellate', 'imperfect' and 'perfect', and (just as Coulter wished) stamens and carpels are referred to as 'microsporophylls' and 'megasporophylls', respectively.[35] Mauseth uses the anthropomorphic language bequeathed by Linnaeus only where it has become established in technical terms of botanical art required by the student of botany, referring, for example, to the 'androecium' (the collective term for the stamens) and the 'gynoecium' (the carpel, including stigma, style and ovary).[36]

It is also possible to find examples of discussions of specific flower structures, usually where the concentration is on only one structure, that avoid all mention of male and female.[37] But far more commonly – to the point that we can say that the practice is hegemonic – plant scientists (and, in their trail, the authors of popular scientific literature and of books for children) refer freely to the stamens/anthers and pistils/ovaries as the male and female sexual organs of the plant, and to male and female flowers and male and female (sporophytic) plants. Reference to the stamens, for example, as the male sexual organ is characteristic of the scientific literature that deals with the reproduction of flowering plants at the highest level of generality, with the least physiological detail, and with no mention of the alternation of generations.[38] It is also characteristic of the popular scientific literature and the children's literature on plants. It is perhaps not surprising to find some bowdlerization in the latter, especially that aimed at primary school-aged children. But the pedagogic constraints of children's literature do not extend to the popular scientific literature aimed at adults, which nevertheless often reproduces about plant sex only what one may find in books intended for, say, nine-year-olds.[39] This is the case even in the plant advocacy literature discussed in Chapter 1, which otherwise aims to introduce the reader to cutting-edge research in plant science and presumes on the part of the reader a willingness to follow more detailed scientific explanations in other areas.[40]

In the scientific literature it is not uncommon for authors to acknowledge, explicitly, that this usage is, strictly speaking, incorrect, before they proceed to employ it.[41] Others acknowledge it implicitly, as when Attenborough writes that 'pollen is conventionally described as being a flower's male cells, and the stamens, which produce it, as the male organs'.[42] But what is perhaps most striking is the widespread and often unremarked use of this 'strictly speaking incorrect' terminology in the field where one might least expect to find it – precisely that of plant sexual reproduction. Where that literature deals with plant cell biology (physiology) and genetics below the level of the flower part, the whole flower or the whole plant, discussion can proceed without any mention of male or female. But where the concern is with populations and mating systems (or sexual systems), plant breeding and evolution, which involves consideration of the gross morphological and quantitative aspects of ovule and pollen production and of floral traits, of individuals and groups of plants, talk of male and female organs, flowers and individuals is commonplace. In other words, it is perhaps where plant sexual 'behaviour' is at issue that the 'old-time' terminology of male and female reasserts itself most insistently.[43]

'The whole case appears to me a very curious one': Darwin on plant sex

However, it was also in this field (plant sexual reproduction) that the most theoretically sophisticated reflection on the concepts used to describe plant sex arose, in the work of the influential New Zealand botanist David G. Lloyd. In a number of papers (some co-authored) on plant sex ratios, breeding systems and sexual strategies in specific plants, Lloyd progressively refines the terminology used to refer to the different 'sexes' of flowers. Lloyd's concerns are botanical and not explicitly philosophical but, throughout, the texts implicitly and explicitly ask one of the central questions of the plant philosophy tradition: 'What do we mean by "male" and "female" in plants?' Lloyd's work may thus be read *as* or at least *through* plant philosophy and doing so yields a result that further complicates both the traditional history of the theory of plant sexuality and the idea that botany becomes properly scientific in understanding plant sex in a 'literal' sense. Lloyd can be read as *constructing* the concepts of 'male', 'female', 'sex' and 'gender' for specifically botanical purposes: that is, concepts more adequate to the specificity of plant life. The meaning of these concepts, for Lloyd, is not read off what is given to the botanical gaze; it is not simply empirically derived. Further, Lloyd's concepts of 'male', 'female', 'sex' and 'gender' are at odds with some, if not all, of the ways that the same words are used in relation to animals. This is because Lloyd's achievement is to have worked these concepts for plants specifically in such a way that they can accommodate both the strictly biological definitions (advocated by Coulter, for example) and the 'old-time' terminology, but without, in the end, staking a claim to have arrived at the final truth of plant sex. If it is nevertheless possible to retain some general biological concepts of 'male' and 'female' and sex, across all relevant kingdoms of life, we will thus find them functioning within a complex field of different concepts of 'male' and 'female' and 'sex' (and 'gender'), giving the lie to the idea that there is anything obvious or simple about sex for any form of life. (We attempt to give a positive philosophical account of that claim in the last part of this book, in the Epilogue.)

An obituary of Lloyd describes him in his early career (1965–75) as 'making observations on natural history in the Darwinian tradition'.[44] However, Charles Darwin's acknowledgement of the complexity of the 'sexual' forms of flowers and his admitted perplexity about these forms can also be understood as the basis of Lloyd's subsequent theoretical reflections on plant sex. In *The Different Form of Flowers on Plants of the Same Species* (1877) Darwin considers certain forms of 'sexual relations' in plants – what are now called 'breeding systems', 'mating systems' or 'sexual systems',[45] that is, the characterization of the species or sub-specific group in terms of the distribution of sexuality across flowers and individuals. He uses some of Linnaeus's basic terminology, distinguishing between monoecious, dioecious, hermaphrodite and polygamous species – that is, those with male and female flowers on the same individual (monoecious), those with male and female flowers on different individuals (dioecious), those with male and female in the same flower (hermaphrodite) and those with male, female and hermaphrodite flowers on the same or different individuals (polygamous). With the meaning of the term 'polygamy' thus restricted, he introduces four other categories: the gynomonoecious (with female and hermaphrodite flowers on the same individual), andromonoecious (with male and hermaphrodite flowers on the same individual); and the gynodioecious (with female and/ or hermaphrodite flowers on different individuals) and androdioecious (with male and/or hermaphrodite flowers on the different individuals).[46] Already we can see that the application of the terms 'male' and 'female' and the presumption of the universality of the male/female binary at the level of the flower and the individual (which is already muddied by the existence of the hermaphrodite: Is the hermaphrodite itself a third sex?) cannot be straightforwardly maintained in the discussion of these new categories. In the gynomonoecious species, for example, each individual is both female and hermaphrodite; in the gynodioecious species each individual is either female or female/hermaphrodite. In these species there seems to be no 'male' at the level of the individual. However, Darwin also introduces an important distinction between form and function, noting that there may be flowers that are ostensibly morphologically hermaphrodite while being impaired in one or other function – for example, morphologically hermaphrodite flowers which in fact produce no seed, thereby being 'in function a male'.[47] This distinction between morphology and function will furnish the basis of Lloyd's conceptual innovations in the thinking of plant sex and, as we shall see, it quite upsets the idea that plant sex is something obvious.

Darwin's main concern in his 1877 text is with the dimorphic and trimorphic sub-groups of the hermaphrodite class. He calls these 'heterostyled plants', that is, plants with two or three forms of hermaphrodite flowers, with a combination of different sized styles (or pistils) and different sized stamens in each. Two main types of these engage Darwin's attention. First, there is *Primula veris* (Cowslip), which has two forms, each with differently sized pistils and stamens. The two forms are known as the 'long-styled' (also called 'pin') plants and the 'short-styled' (also called 'thrum') plants (see Figure 5.3). In the discussion of this species Darwin recounts that, in asking himself about the 'meaning' of the existence of these two forms, he first thought that the species must be evolving towards dioecy and presupposed 'that the long-styled plants, with their longer pistils, rougher stigmas, and smaller pollen grains, were more feminine in nature, and would produce more seed' and that the short-styled were more 'masculine'. But, contrary to his expectations, collections of seeds from both showed the latter to be the more fertile, which for him meant that the short-styled proved to be more feminine in nature.[48] It is notable that he here introduces (and continues to maintain) a terminological distinction between 'male' and 'female' and 'masculine' and 'feminine'.

In further experiments Darwin discovered that long-styled flowers pollinated with pollen from the short-styled form were more fertile (i.e. they produced more seeds) than those self-pollinated or pollinated from other long-styled plants. Conversely, the short-styled form was most fertile with pollination from the long-styled form.[49] The grades of fertility in individuals could range from zero (completely infertile) to

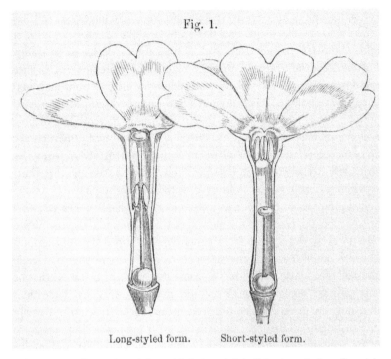

Fig. 1.

Long-styled form.   Short-styled form.

FIGURE 5.3 Illustration of dimorphic heterostyly in *Primula veris*, from Darwin's *The Different Forms of Flowers*. We can see that both the style and the stamens are of different sizes in the two forms. By permission of the Linnean Society of London.

'perfect' or 'full' fertility, with all stages between. With perfect fertility as the ideal, Darwin then reconceives the sexual nature of the dimorphic heterostyled species:

> The individual plants of the present species [*Primula veris*]... are divided into two sets or bodies, which cannot be called distinct sexes, for both are hermaphrodites; yet they are to a certain extent sexually distinct, for they require reciprocal union for perfect fertility. As quadrupeds are divided into nearly equal bodies of different sexes, so here we have two bodies, approximately equal in number, differing in their sexual powers and related to each other like males and females.[50]

As the use of the animal analogy betrays, Darwin is searching here for a binary that is not in fact supported by the facts as he finds them.

For, according to his own lights, those who are 'like males' can fertilize others who are 'like males', and those who are 'like females' can fertilize others who are 'like females', even if they are less fertile in so doing. But the important point is that the male and female *function* is once again being conceived – albeit confusedly – as distinct from male and female morphology. This function is also expressed in terms of 'sexual powers'. Later, in relation to another dimorphic heterostyled species, *Linum grandiflorum* (known as Flowering Flax, amongst other names), Darwin's experiments seemed to show that the long-styled form was absolutely sterile when pollinated with own-form pollen and the short-styled less fertile when pollinated with own-form pollen.[51] As the pollen grains from both forms were undistinguishable under Darwin's microscope, the nature of the 'power' of the stigmas to cause the pollen tube to grow from other-form pollen was to him deeply mysterious. But it is established that the two pollens and the two stigmas clearly 'mutually recognise each other by some means',[52] and that it is what he later calls the 'sexual elements' or 'reproductive powers' that determines this, not 'any general difference in constitution or structure'.[53] Darwin thus distinguishes the male and female 'fertilizing power' from the form of the male and female flower or male and female individual, evoking the Aristotelian distinctions between male and female principles, male and female organs and male and female individuals – or at least highlighting, once again, the need to revive the question of what is meant by 'male' and 'female' in each case.

The second type of heterostyled plants that Darwin discusses complicates things even further. *Lythrum salicaria* (Purple Loosestrife) is trimorphic (see Figure 5.4). The manner of fertilization of this plant, according to Darwin, offers 'a more remarkable case than can be found in any other plant or animal'. In his own description:

> The pistil of each form differs from that in either of the other forms, and in each there are two sets of stamens different in appearance and function. But one set of stamens in each form corresponds with a set in one of the other two forms. Altogether this one species includes three females or female organs [long-styled, mid-styled and short-styled] and three sets of male organs [combinations of 'longest', 'mid-length' and 'shortest'], all as distinct from one another as if they belonged to different species; and if smaller functional differences are considered, there are five distinct sets of males.[54]

CHAP. IV. LYTHRUM SALICARIA. 139

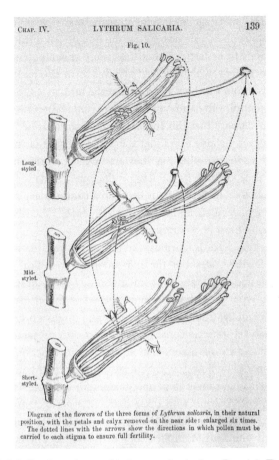

Fig. 10.

Long-styled

Mid-styled.

Short-styled.

Diagram of the flowers of the three forms of *Lythrum salicaria*, in their natural position, with the petals and calyx removed on the near side: enlarged six times. The dotted lines with the arrows show the directions in which pollen must be carried to each stigma to ensure full fertility.

FIGURE 5.4 The three forms of *Lythrum salicaria*, from Darwin's *The Different Forms of Flowers*. By permission of the Linnean Society of London.

Darwin's experiments showed him that the greatest fertility is achieved when pollen from the longest stamens pollinate the longest pistil, when pollen from the mid-length stamens pollinate the mid-length pistil and when pollen from the shortest stamens pollinate the shortest pistil – these are what he calls the 'legitimate unions'.[55] However, most of the 'illegitimate' unions also resulted in seed, albeit in much reduced number. The fact that these experiments required Darwin to make 'eighteen distinct unions in order to ascertain the relative fertilizing power of the three forms' shows 'the extraordinary complexity of the reproductive system of this plant' for which, he says, 'nature has ordained a most complex marriage-arrangement, namely, a triple union between three hermaphrodites'.[56] He also found the mid-styled form to have the highest capacity to be fertilized but the lowest pollen potency. He cites previous experiments which showed that this form also has the smallest pollen grains. From this he concludes that the mid-styled form 'appears to be highly feminine in nature' and, as with *Primula veris*, his speculative instinct is to suppose that this form is tending towards becoming fully female. However, he also notes that the female organs of this form are already fully fertile, meaning that they are in effect already fully female. At the same time, because its stamens produce an abundance of pollen, it is not possible to consider them as being in a rudimentary condition. It is both fully female, then, and at least partially male. 'The whole case', he says, 'appears to me a very curious one'.[57]

It is clear that what is 'curious', for Darwin, is not the mere fact of differential fertility rates but the impossibility of a clear distinction between male and female and the combinations of different sexual organs of the same 'type', such that the same species is found 'bearing three females, different in structure and function, and three or even five sets... of males; each set likewise consisting of half-a-dozen [stamens], which likewise differ from one another in structure and function'.[58] Multiple 'sexual organs' in a flower is of course nothing unusual; almost all flowering plants have more than one stamen, for example. But *Lythrum salicaria* presents Darwin with three different *types* of females and up to five different *types* of males. The basic sex duality of male-female is here replaced with a sex multiplicity of female1-female2-female3-male1-male2-male3-male4-male5. Moreover, some females are more feminine than others, and some males more masculine, suggesting not a sex duality but a sex continuum, even at the level of the organ. Any straightforward transfer of the sex duality male-female from organ to flower to individual is thus thoroughly disrupted. The differential fertility of dimorphic heterostyled plants, some of which are infertile with own-form pollen, allows Darwin to *project* a sex duality on them,[59] but *Lythrum salicaria* resists even this and forces the question of what we mean by 'male' and 'female' at different levels (sexual organ, flower, individual) and of the relationship between these levels. Basically, the conflation of the different meanings of male and female that is both the presupposition and the result of Linnaeus's sexual system – much of the terminology of which Darwin assumes – here begs to be disaggregated and rethought. The application of the basic zoological model of sex to plants – in which one does *not* ask 'What are "male" and "female"?' – is revealed to be not only inadequate but also confusing (both conceptually and terminologically) in trying to understand the specificity of vegetal sex. Plants appear as sexual oddities as compared with animals rather than as manifesting 'normal' forms of a specifically vegetal sexuality. The plants need, instead, a plant model. This is what David Lloyd attempted to provide.

The plant model of gender: David G. Lloyd and the sex/gender distinction in plants

In the earliest of the essays with which we are concerned here, one feature of Lloyd's work across the next decade already emerges: a frustration with what he will later call 'typological' approaches to the description of sexual systems (also called 'breeding systems' or 'mating systems') in plants. A 'sexual system', as we noted earlier, is the characterization of the species or sub-specific group in terms of the distribution of sexuality, or 'sex expression',[60] across flowers and individuals. The idea of 'sex expression' is already ambiguous, referring both to the sex expressed in any particular individual or flower and to the characteristic forms of such expressions in any given species. It is the latter which is a sexual system. Lloyd criticizes the traditional mode of identification and definition of such systems, that is, the traditional morphological *method* of identification and the typological *result* of that method, although the form of the result is already contained in the method (the search for morphological types or a distribution of types). The probably unwitting presupposition of this is the idea of a fixed set of possible sexes and possible sexual systems. This was already being questioned in Darwin's *The Different Forms of Flowers on Plants of the Same Species*, where, as previously discussed, he implicitly raises the idea of a sex continuum, but the idea was still strong enough in 1972 for Lloyd to feel the need to argue against it. This is obviously not because any plant scientist in fact held an explicitly static view (a significant amount of the literature on sexual systems is concerned with, for example, the evolution of dioecy from hermaphroditism); it concerns the implicit effects of the continued use of a terminology derived from a Linnean-style typological impulse and the lack of any determined alternative principle and method for the identification of sexes and sexual systems in plants. In work from 1972 to 1980 Lloyd progressively offered such an alternative.

The identification of sexual systems is obviously bound up with the identification or definition of sexes. In introducing an alternative conception of plant sex, Lloyd switches, quite deliberately it seems, to the term 'gender', with some essays effectively reserving 'sex' for the older morphological conception.[61] However, he also frequently speaks of the older model in terms of 'morphological' or 'phenotypic' gender, which he associates, somewhat mysteriously, with 'verbal descriptions'. The identification and description of the 'phenotypic gender' of an individual plant is 'based on the reproductive characteristics of that plant alone – counts of the percentage of ovuliferous, seminiferous, or polleniferous flowers, for example'.[62] Thus the identification of 'phenotypic gender' is centred on the individual plant, in isolation from others of its species, with the visible sex 'organs' playing the primary role: in effect, organ morphology is plant sex. This is the presupposition of the 'old-time' terminology that, while strictly speaking incorrect, is still widely used.

Lloyd does not object to the old-time terminology on the basis of its being strictly speaking incorrect; so what, for him, is the problem with this morphological conception of plant sex? According to Lloyd it is that it tends to presuppose neat divisions between sexual systems and the sexual expression of individual flowers and plants, such that exceptions to the neat divisions (e.g. the appearance of hermaphroditic flowers in dioecious populations, pollen-bearing flowers on otherwise 'female' plants, polleniferous parts in otherwise 'female' flowers and so on) are 'dismissed as unimportant anomalies, irregularities, or "intersexes"': that is, as freaks.[63] Just as the mammal hermaphrodite or intersexed individual is seen as an anomaly or irregularity, deviating from the strict male-female dichotomy, so the polleniferous female plant is seen as an irregularity deviating from the strict, morphologically derived male-female-hermaphrodite trichotomy. But what if the freak is normal? What if the neat typology is the problem?

Substituting a functional criterion for the morphological one, Lloyd proposes a set of definitions that can accommodate all irregularities. However, they do not do this by multiplying the number of recognized gender classes or sexual systems.[64] Lloyd's is not an MFH+ approach. Instead, Lloyd's functional criterion for plant gender maintains the traditional or old-time categories of male and female but redefines them in an inclusive manner (there are no freaks; no plant is an exception) that allows for – indeed expects – the gender expression of some individuals and of populations to change over time. It is a generous concept of gender. Although Lloyd begins to put a functional concept of plant gender to work as early as 1972, the fullest theoretical expression of this conceptual innovation is found in two essays from 1980. These lay out two main aspects to the functional concept of plant gender.

First, plant gender is quantitative. The open system of plant growth and its modular form mean that the 'sexual organs' are typically repeated many times in any given individual. Whether the 'descriptive units of sex' are sporophylls, sporangia or gametes, 'the relative proportions of male and female organs may vary over the entire range between the all-male and all-female extremes. The intermediate conditions are often fully functional. Detailed descriptions of plant gender should therefore be quantitative'.[65] The functional, quantitative conception of plant gender measures the relative paternal and maternal investments of a plant, calculating 'the proportions of a plant's genes that are most likely to be transmitted through pollen (its maleness) or through its own seeds (its femaleness), based on expenditure on androecia and gynoecia'.[66] Calculating the gender of a plant – its functional maleness or functional femaleness – is based on pollen, ovule or seed counts, or on the number of flowers bearing pollen, ovules or seeds. It thus still relies on the observational methods of the natural history tradition and of Darwinian botany. It further accepts that all calculation of plant gender will be an estimation.[67] Sometimes, notably in the case of dioecious populations with little internal variation, the functional gender will match the morphological sex of the traditional method: plants bearing exclusively gynoecia are female; those bearing exclusively androecia are male.[68] But often, especially in the case of gynodioecious populations where the morphological approach identifies females and hermaphrodites, the functional designation of gender will be at odds with the morphological designation. In a gynodioecious population those plants with exclusively gynoecia are functionally female because the plant's genes are obviously most likely to be transmitted through the maternal function. In the same population a plant with both gynoecia and androecia is more likely to transmit its genes through the paternal function, so it is described as functionally male,[69] even if it has functional gynoecia and even if it turns out that any given individual or flower *in fact* performs a female function, that is, is pollinated and fertilized. This is perhaps the most counterintuitive conclusion from the point of view of the traditional, morphological conception of plant sex: males may have functional female organs and males may be fertilized by other males and bear fruit and seed.[70] Similarly, in an androdioecious population, functional females may and probably will have functional androecia and fertilize other functional females. There is no contradiction, therefore, in the idea that some males perform a female function or vice versa: 'Maleness and femaleness in plants are recognized as quantitative phenomena, not absolute distinctions'.[71] There is a 'gradation rather than a discontinuity in gender values'.[72]

The second main aspect of Lloyd's functional definition of gender is that 'it has no meaning for single plants considered apart from the population in which they would interbreed'.[73] Whereas the morphological (phenotypic) conception of gender *only* considers single plants (or flowers), the functional conception of gender 'considers a plant's sexual performance in relation to that of the other individuals in the population'.[74] Whether any plant with both gynoecia and androecia will be male, female or 'co-sexual'[75] depends on the other plants in that population. The morphology of gynoecia and androecia alone cannot determine it except in the case of strictly unisexual populations (where individuals can only produce either male or female gametes, never both), and even then occasional 'hermaphrodite' flowers of plants will still be classed as either male or female (whereas in a monomorphic population with only such flowers, they would be co-sexual).

Together, these two aspects of Lloyd's functional concept of gender mean that the identification of any plant or flower (or any 'morph')[76] as 'male', 'female' or 'co-sexual' is not a static identification of the innate sexual being or nature of any given individual or flower but a snapshot within a dynamic population, especially where – as is common – the different sexual systems and genders shade into each other.[77] As Lloyd says in 1979, 'the terms "cosex", "male", and "female"… refer to the average performance of a morph and do not specify the limits of behaviour of the individuals of that morph'.[78] As this concept of gender is based on potential, not actual function, the identification of plant gender is also in many cases speculative. Lloyd acknowledges that the complete estimate of the functional gender of plants – which requires maternal and paternal fitness to be measured through gamete production, fertilization, seed maturation and dispersal, and the growth of offspring to adulthood – is 'a formidable or even impossible task'.[79] Nevertheless, incomplete estimates are valuable in determining the likely relative success of a plant as a maternal or paternal parent. Again, this stresses that it is the sexual strategies of populations of a given species that determine the identification of the gender of any individual plant, marking a significant break from the individual-centred morphological approach and definitively casting aside the idea that there is anything obvious about plant sex.

The relational epistemology of vegetal sex

What is the relation of Lloyd's theoretical work on the concept of plant gender to the terminological disputes about plant sex that we discussed earlier in this chapter? The positions in those disputes seemed to be polarized between two alternatives: on the one hand, an insistence on the technically correct restriction of the idea of plant sex to gametes and gametophytes (as in Coulter); on the other, a plea for the continued, common-sensical and analogical use of the old-time, 'untechnical' terminology of male and female to refer to the parts of flowers and to individual plants, that is, extended metonymically to the sporophyte generation (as in Bailey). But Lloyd's work does not accord with either of these alternatives; rather, it supersedes them in proposing new, specifically vegetal definitions of male and female and distinguishing between the concepts of sex and plant gender.

The quantitative concept of plant *gender* privileges the population over the individual and function over morphology and in so doing redefines the concepts of male and female *for plants*. This does not deny the basis of the definitions of male and female in the gamete, but it does refuse the progression of definition, implicit in the morphological individual conception of plant sex, that sees the gamete as defining the organ and the organ as defining the sex of the flower and/or plant. Lloyd's is a relational conception of gender that thinks gametes (represented in investments in gynoecia and androecia) in the individual plant or flower *only* insofar as they are subordinated to their potential distribution within a population. The fact of the potential or actual production of female gametes does not necessarily render a flower or plant female or hermaphrodite; rather, the total maleness or femaleness of a population determines the gender of the individual. Similarly, the identification of a morph as male or female attributes to it an affinity with a population class; it does not specify its individual nature.

As we have said, Lloyd's use of the term 'gender' seems deliberate, marking the conceptual innovation of his work. And although he continues to use words and phrases such as 'sex expression', 'sex ratios', 'cosexual', 'sexual class' and so on, the theoretical thrust of the work is towards a distinction between functional plant *gender* and the traditional understanding of morphological plant *sex*.[80] Lloyd's redefinitions of male and female in relation to the former mean the identification of a plant's gender may well conflict with the identification of its sex, traditionally understood: 'It is frequently assumed that plants in one population and with the same general sex form are identical in their gender, but different plants of the same general phenotype can vary considerably in their precise gender.'[81] But if Lloyd's position is clearly that functional gender is superior to phenotypic sex as a measure of sexual behaviour in plants,[82] what is the status of the idea of morphological plant sex in his work? Or what can we now say about that idea?

Lloyd does not reject the idea of morphological plant sex entirely but sees it as an indicator of last resort when the information that would enable one to calculate plant gender is unavailable. Given the complexity and time investment required to calculate functional gender, Lloyd accepts that designations of morphological plant sex will sometimes still be used and indeed are useful, just as the 'typological' distinction between sexual systems, although crude, is useful:

> Gynodioecy is not a uniform breeding system but a wide range of conditions… [it] nevertheless remains a useful term to describe a class of plants with distinctive properties in common… neither the presence of a continuum in sex functions nor the use of quantitative measures to describe these functions removes the need for a simple verbal description of such populations.[83]

The characterization of the morphological sex categories as 'verbal description', 'verbal label[s]' or 'verbal terms'[84] contrasts them with the theoretically determined *concept* of functional plant gender and of the *concepts* of male and female associated with it. That is, whereas plant 'gender' and Lloyd's redefinitions of male and female *for plants* are basic *concepts* grounding his investigation of plant sexual behaviour, the terms of morphological plant sex are just *words* applied to plants for convenience's sake. Moreover, they are words derived from an animal model. Thus Bailey's plea for the continued use of the old-time untechnical vocabulary is respected, but it is clearer what this use involves: a pragmatic compromise which is very far from the 'literal' depiction of plant sex hailed by the traditional historians of botany, being, on the contrary, 'a manner of speaking' about plants borrowed from another form of life.

Lloyd's concept of plant gender, on the other hand, and the definitions of male and female for plants that follow from it, attempts to move beyond the animal model to propose specifically vegetal ways of thinking about plant sexuality. Lloyd, of course, is only concerned with the consequences of this reconceptualization for the plant sciences, and particularly for a more adequate approach to the understanding of sexual reproductive strategies in plant populations. But the work can also legitimately be interpreted as a series of propositions for plant philosophy concerning vegetal sex, propositions that also have implications for the way that we understand sex more generally (as we will discuss further in the Epilogue).

First, what does Lloyd's reconceptualization of plant gender allow us to *know* about vegetal sex? Lloyd utilizes the traditional categories of the animal model: male and female. But the redefinition of these categories in functional, quantitative terms breaks the presumption of the animal model which, we can now see, operates according to a kind of ontic *emboîtement*: the gamete is the smallest male or female unit, nesting (at least at first) within the larger male or female unit of the organ, which is itself nested in or part of the largest male or female unit, the individual. In the animal model the direction of identification moves from the gamete to the organ to the individual, or the determining criterion of the identification of the largest male and female unit is based in its ontological relation to the smallest (the being of the smallest determines the being of the largest). Nevertheless, the logic of the individual prevails at each level of the ontological determination of sex: a gamete, an organ, an individual male or female animal. The animal model is fundamentally a system of ontological metonymy controlled by the logical priority of individuality.

Lloyd's concept of plant gender, though it has its biological basis in the idea of the gamete, is, on the other hand, a concept that is conceived in terms of multiplicity. The direction of identification moves from the population (a dynamic multiplicity) and the potential distribution of the multiplicity of gametes within it, to assign a gender *value* to an individual. As a value, plant gender is not an attribute of a plant subject but its coordinates, its current gender position within a population. This plant model is fundamentally a system of mathematical coordination controlled by the logical priority of multiplicity (but not collectivity). It is not an ontology of plant sex; there is no ontology of plant sex. It is the mobilization of a relational epistemology where what is known is not the literal truth of a plant's sex but the gender value of its current allocation in relation to a population.

Lloyd is clear that a gender value is an estimate and that the identification of the 'effective' (i.e. actual) gender of a plant may be impossible. But, philosophically, this conception of plant gender is not proposed *faute de mieux*. It is not a temporary stand-in while we wait for the truth of plant sex to be delivered up in the form of an ontology that we can recognize. It is more fundamentally an acknowledgement of the alterity of plant life and the alterity of plant sexuality. Lloyd's formulas for the measurement of plant gender yield estimates that acknowledge, in effect, that we do not know the 'measure' of plant sex because we have not yet rid ourselves of the conceptual instruments attuned to the key of animal being.

But are these instruments out of tune, even for us?

6 ARE WE FAMILY? THE MOTHER TREE AND OTHER HUMANS

As we have noted, the zoological model and the analogy with animals that provided the framework and the inspiration for much of the botanical tradition until well into the eighteenth century increasingly came to be perceived as an obstacle to a properly scientific botany. The identification of the Aristotelian tradition in botany with the zoological model and the analogy with animals means that traditional historians of botany tend to see the rejection of the animal model and the rejection of philosophy as two sides of the same coin. The Aristotelian tradition of plant philosophy is also associated with the denial of the idea that plants have male and female, but this is a mistake based on the presumption that 'male' and 'female' refer primarily to individuals or parts. As we have shown, the central feature of the discussion of plant sex in the Aristotelian botanical tradition is the recognition that the question 'Do plants have male and female?' requires an answer to the prior, more general philosophical question 'What are "male" and "female?"' Attention to this question led to an acknowledgement of the different possible senses of the words, an insistence on distinguishing between them and attempts to specify the sense(s) in which plants do or do not have male and female. Of course, Aristotle's distinction between the metaphysical principles of male and female (which he understood as united in plants), male and female individuals and male and female parts or organs is not straightforwardly going to provide the basis for any plant science today. But the philosophical impulse behind his question, the continued need to ask it and to make relevant distinctions, does still have a significant role to play and is of philosophical interest beyond botany.

We have also noted that the zoological model and the analogy with animals in fact survive both in much of the terminology still used in the plant sciences and in popular ways of speaking about plants today. This is particularly the case in discussions of plant sex and plant sexual reproduction. As we argued in the previous chapter, the discovery of the unique, dibiontic life cycle (or alternation of generations) in plants revived the Aristotelian question of what we mean by 'male' and 'female' in plants. As a result, it is generally recognized that it is not strictly speaking correct to refer to the flowers, flower parts or individual sporophyte plants as 'male' and 'female', and indeed it is recognized that this practice is a remnant of the zoological model and its presumptions – the presumptions that one will find sexual organs and sexed individuals bearing those organs, just as is common in the animal realm.[1] The widespread – if largely implicit – agreement to continue, nevertheless, to use this 'untechnical' vocabulary thus shows that the analogy with animals is still a significant feature of discussions of plant sex today in scientific journals, in popular science and in children's books. Furthermore, and contrary to the expectations of traditional histories of botany, it is where discussions of plant sex are closest to plant philosophy – in the theoretical work of David G. Lloyd, for example – that the limitations of and problems with the animal model are addressed. That is, the most serious attempts in the plant sciences to come to terms with the specificity of plant sex – with what it means to talk about 'male' and 'female' in plants – arise in the continuation, not the repression, of the philosophical impulse in botany.

In the plant sciences discussion of mycorrhizae have also successfully engaged the public imagination. It is striking that other kinds of zoocentric – and more specifically anthropomorphic – analogies again play a central explanatory role in this. The nature, and extent, of the underground mycorrhizal networks in forests is becoming clearer. The understanding of the role of these networks in connecting plants – especially trees – and facilitating nutrient and water exchange between them is revealing a specifically fungal and phytological mode of inter-existence, the recognition of which could revolutionize forest management and which may well be exceptionally important in understanding climate change and in developing strategies to combat its effects. Couched in the terms of human sociality and human kin relationships, it also seems to offer attractive models for human co-existence (intra-human, between species and across realms). The question of the relationship between the findings of this Western scientific work and much older and more deeply socially embedded indigenous knowledges is complicated and needs to be approached carefully.[2] But the perceived 'authenticity' of Western scientific work when it seems to echo aspects of indigenous wisdoms is no doubt part of its appeal to the alienated (perhaps anti-capitalist, eco-critical or simply jaded, exhausted and proletarianized) Western subject. Probably the most influential and enthusiastically received example of this is Suzanne Simard's work on the ecology of the forest and its 'Mother Trees'.

Simard's work focuses on both microscopic and macroscopic phenomena. Indeed, it is partly about the relationship between the very tiny (e.g. carbon molecules) and the vast (whole forests and their place in global ecosystems). Although much of it is concerned with the conditions under which seedlings thrive (or not) and relationships between parent trees and their offspring, it is not about sexual plant reproduction per se. But we can nevertheless ask: What is the relationship between the zoocentric terminology of discussions of plant sexual reproduction – focusing especially on the plant 'embryo' – and the understanding of vegetal reproduction, growth and existence in the anthropomorphic terms of human kinship relations? What is the significance of the shift towards an understanding of vegetal relations via a specifically anthropomorphic analogy, as in Simard's work? These questions take us back to the ambit of the plant advocacy literature discussed in Chapter 1, but also beyond it, to its intersection with attempts to marry Western science with traditions of indigenous wisdom and with the speculative philosophical anthropology (or anthropological philosophy) to be found in the work of Eduardo Viveiros de Castro. This may seem to take us far from plant philosophy, but in fact Viveiros de Castro's work allows us to raise the problem of anthropomorphism in discussions of plants in a new way. In this final chapter, then, we will examine the idea of the Mother Tree in the context of what we will call the 'maternal botanical imaginary', in relation to attempts to talk about the possible connections between indigenous wisdom and Western science and philosophy and through the critical challenge posed by Viveiros de Castro's anthropo-philosophy. Is the idea of the Mother Tree merely another anthropomorphic trope, and one which, moreover, pays insufficient attention to the conservative and even reactionary aspects of its exclusive association of nurture with the maternal? Or are we, rather, witnessing an a-human and specifically vegetal transvaluation of the human, and perhaps of the maternal too? Finally, can we pursue Viveiros de Castro's philosophical opening in such a way that will allow us to rethink the generality of the generic concept of sex?

The plant embryo

In his review of the history of discussions of the plant embryo, Hans Werner Ingensiep notes that what he calls the 'zoomorphic' terminology and approach of early botany meant that plant development was imagined in the same way as animal (including human) development.[3] As we saw in Chapter 2, the analogy between the animal egg and the plant seed was already common in Western antiquity, and Aristotle characterized the seed as a 'fetation' (*kuēma*).[4] Reference to the plant 'fetation' in the Aristotelian sense of the product of a commingling of the male and female principles is then common in the plant philosophy tradition up until Cesalpino, who first refers to the plant 'foetus' in *De plantis* in the context of the opening section on analogies between animals and plants.[5] Ingensiep notes that this analogy allowed botanists to anticipate the as-yet-invisible stages in early plant development, before the invention of the microscope.[6] But the influence of the analogy is also clear in the new plant anatomy, inaugurated by Grew and Malpighi, facilitated by microscopic observation. Grew calls the 'seed case' the 'uterus' and writes of seed development: 'in the greater number of *Seeds*, is formed a true *Bud*, consisting of perfect *Leaves*; different from those, which grow upon the *Stalk*, only in *Bigness*; and so far in *Shape*, as the same *Parts* of an Animal *Foetus*, in its several ages in the *Womb*'.[7] Malpighi, though, was the first to offer a detailed anatomical description of the development of the foetus within the seed, via a more developed analogy with animals. With the almond as his example (see Figure 6.1), the parts of the various stages of fruit, seed and plantlet are named after the allegedly corresponding parts in animals: for example in Malpighi's figure 231 'A' is 'uterus'; in his figure 233 'I' is 'amnion' (the name for the membrane covering the animal foetus, creating the sac that fills with fluid), 'L' is 'umbilicus' and 'M' is 'chorion' (the name for the outer membrane of the embryo sac in animals). 'H' is the 'foetus'.[8]

The identification of the plant embryo within the seed was quite compatible with a range of theories of generation (e.g. animaculism and preformationism) and with arguments both for and against the sexuality of plants. The use of the term 'embryo' in botany thus functioned, as Ingensiep puts it, as a 'conceptual cluster for describing and deciphering shapes in early stages of plant development'.[9] It is not the result of a discovery, the discovery that the plant develops from the embryo just like the animal. It is part of a process of the transformation of an analogy with animals

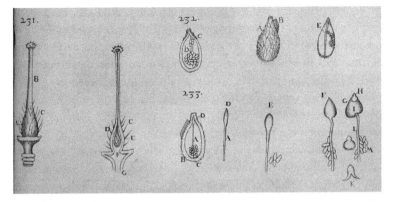

FIGURE 6.1 Malpighi, *Anatome Plantarum*, Plate XXXVII, figures 231–3. By permission of the Linnean Society of London.

into an identification of parts and the adoption of a scientific terminology, borrowed from studies of animal development, into botany. By the end of the eighteenth century, and thanks to the influence of the German botanist Joseph Gärtner, the term was widely accepted as 'correct'.[10]

By the mid-nineteenth century 'amnion' was replaced with 'embryo sac'[11] and the plant embryo had acquired a plant placenta. Again, the passage from analogy to identity can be traced in the development of the idea of the plant placenta. In 1797 Robert Hooper was comparing (analogically) both the plant cotyledons in seeds and the small vessels in tree buds to the animal placenta.[12] He does not say that the plant *has* a placenta. However, by 1826 a part of the plant is described that is not just analogous to the animal placenta, it *is* the plant placenta. Again, this is not the result of the discovery that the plant has a placenta just like the animal but the decision to borrow the word from zoology to name a specific (albeit not initially precisely determined) part of the plant.[13] In John Lindley Thomas Moore's botanical dictionary of 1866 the plant placenta is defined as 'The place or part on which ovules originate', which is pretty much how it is still defined today.[14]

As has often been noted, the use of the same terminology to name different structures in animals and plants can cause confusion; 'ovary' and 'placenta' are cases in point. The animal placenta develops from the zygote after it has developed into the early embryo (the blastocyst), from the outer layer of cells, whereas the plant placenta develops as part of the ovary regardless of whether fertilization occurs (the plant placenta grows independently of the zygote and indeed in its absence). The animal ovary is the site of the development of ova (the female gametes, or their precursors, oocytes), whereas the plant ovary is the site of the development of the ovules, in which the megaspores develop that subsequently develop into the megagametophyte which produces the megagamete.[15] The plant embryo grows within the plant ovary, whereas the animal embryo obviously grows within the uterus (so we understand why Malpighi called the plant 'ovary' the 'uterus'). German botanical terminology recognizes the phytological specificity of the plant 'embryo', 'embryo sac', 'ovary' and 'ovule' by calling them *Keim*, *Keimsac*, *Fruchtnoten* and *Samenanlage* respectively.[16] French botanical terminology, as Hallé points out, is in danger of even more confusion than the English as it does not distinguish between plant ovule and animal ovum, using the word '*ovule*' for both.[17]

The maternal botanical imaginary

In English the constellation of 'embryo', 'embryo sac', 'ovary' and 'ovule', most of which are specifically female structures, gives rise to a maternal botanical imaginary. This 'imaginary' is to be understood in the sense proposed by Michèle le Doeuff. In this sense 'images' (including linguistic images) are not radically heterogeneous to objective scientific work, but neither are they absorbed into the 'conceptualized problematic' of an area of science, such that their meaning is simply 'congruent with the theoretical results' which they only translate or illustrate. Rather, the imaginary is part of the theoretical enterprise, an element within it, not a 'dross coming from elsewhere, or a duplicate, serviceable to the reader's deficient culture yet dispensable'.[18] Le Doeuff writes that 'imagery copes with problems posed by the theoretical enterprise itself', but this does not mean that it solves them: 'the imaginary which is present in theoretical texts stands in a relation of solidarity with the theoretical enterprise itself (and with its troubles)' – it is 'at work'.[19] In speaking of a maternal botanical imaginary we mean, then, that the invocation of the mother and her maternal care, evoked by the conceptual constellation that clusters around the plant embryo, is to be thought (for good or ill) as part of the theoretical understanding of the plant, not merely a way of dressing it up for the people.

When we pay attention to the maternal botanical imaginary, we find that there are mothers and embryos all the way down. For example, in both the male and the female sporophyte, sporogenesis is said to begin from mother cells (the pollen mother cells – therefore the mother in the father – and the megaspore mother cell).[20] Across all areas of biology embryonic (and other) stem cells are also called 'mother cells', and in mitotic division the dividing cell is the 'mother', her genetically identical offspring her 'daughters'.[21] In this respect botany follows standard cytological terminology (albeit cytology is the younger discourse). In addition, however – and no doubt because of the internalization of the megagametophyte in the angiosperm sporophyte – each stage of plant sexual reproduction seems to evoke the mother-embryo relationship. Meeuse and Morris compare the place of the female gamete (the egg cell) within the ovule as 'a situation analogous to the maternal care of the mammal embryo', held in her 'protective custody', but the ovule itself is also 'held in the protective embrace of the mother plant'.[22] Hallé compares the female gametophyte, reduced to the embryo sac, to 'an embryo in the belly of its mother'.[23] This is an image that in E. J. H. Corner's interpretation of events means that the dibiontic life cycle of the plant can effectively be seen as contracted into the monobiontic animal form such that 'the forest tree reproduces viviparously: maternal care is botanical, as well as zoological'.[24] In discussion at the level of the mature sporophyte and at the level of the population, the mother plant is both the fruiting plant (so the plant that does actually contain the plant embryo within it; although it might also be a 'father', that is, have supplied pollen) and the parent of any vegetative (non-sexually reproduced) offspring. It is to that extent the genealogical or genetic mother plant. Population studies, especially, make frequent reference to maternal and paternal fitness – again, speaking genealogically, in terms of the passing on of genetic material. But as we can see from the quotations above, the genealogical mother shades easily into the nurturing mother – or the idea of genetic parenthood shades easily into the idea of maternal care.[25] This maternal botanical imaginary is evoked even where the more explicitly genetic or genealogical idea of the parent is involved, especially in the idea of 'maternal investment' in plants.[26]

The Mother Tree

In the relevant popular scientific literature this aspect of the maternal botanical imaginary – the idea of maternal care – comes to dominate, as the mother's role shifts from gestation to upbringing. It is here that the imaginary emerges most clearly, not least because of the explicit acknowledgement of it and its role in scientific practice. The scientific work on which this is based comes mainly from Suzanne Simard and her co-workers. All of Simard's work concerns the total ecology of the forest: relations between organisms (including trees and plants of the same species, different species, fungi, bacteria, insects and birds) and their relations to the local and global climate, to the soil and rock. Her research was first conducted in the context of the forestry practice of clear-cutting – razing managed pine forests and replanting seedlings in areas made and kept free from other plants as much as possible, including with the use of herbicides. Clear-cutting presupposes that different species will inevitably and only compete for soil resources and light, and that freeing fir trees from the competition (which is classed as 'weeds') will result in more and better wood production. Simard established that, on the contrary, fir seedlings flourished when they were able to become established with mycorrhizal networks (seedlings in experiments fared much better when planted in old growth soil that already contained fungal organisms, and not in those where herbicides had killed them) and, perhaps more surprisingly, when planted with birch seedlings. Although transfer of elements via connecting mycelia had been demonstrated in laboratory experiments, Simard's research team was the first to prove that the same happened in the field in the context of 'source-sink' relationships – that is, transfer from carbon-richer to carbon-poorer organisms. A famous letter in *Nature* in 1997 reported results of field experiments showing bidirectional carbon transfer and then, after two years, net carbon transfer from Paper birch to Douglas fir seedlings, increasing according to the extent to which the fir seedlings were planted in shade.[27] Subsequent research showed that the transfer of carbon between species was variable, depending on season and that species benefiting from net transfer while young could themselves become a transfer source as they outgrew their earlier donors.[28]

Simard's work also focuses on the mycorrhizal networks themselves, unearthing, mapping and modelling the underground networks, leading her to conceive of the forest as a complex, dynamic, adaptive, intelligent system,[29] an understanding of which is crucial in developing strategies to manage or mitigate climate change.[30] To this extent she is increasingly associated with the research in plant behaviour and intelligence that we discussed in Chapter 1.[31] But Simard is probably best known for her popularization of the idea of the Mother Tree. The scientific basis for this is the results of research which seem to show, for example, that seedling survival in drought conditions is maximized when those seedlings are able to access mycorrhizal networks connecting them with older, more established (and thus less vulnerable) trees.[32] One study concludes that the mycorrhizal symbiosis is not just between two or more organisms but is a complex assemblage of fungal and plant individuals spanning multiple generations. All 'individuals' are thus nodes in the network, but the same study shows that within this assemblage the degree of connectivity ('node degree') of an individual tends to correlate with its size and age. The tree with the highest 'node degree' in one cluster was found to be directly linked to forty-seven other trees:

> This corresponded to 84% of potential linkages between this tree and all other trees encountered in the plot, and was three times higher than the mean node degree among trees. The influence of this tree as a network component, despite only a portion of its roots being sampled (the bole [trunk] of the tree was located 4.2 m outside the plot), suggests that it would be an even stronger hub at a larger spatial scale.[33]

Although Simard does not seem to use the phrase in her work published in scientific journals, this super-connected tree is what she elsewhere calls a 'Mother Tree'. Indeed, since 2015 Simard has been leading The Mother Tree Project, 'investigating forest renewal practices that will protect biodiversity, carbon storage and forest regeneration as climate changes',[34] and there can be no doubt but that this is exceptionally important research.

The place of the notion of the Mother Tree in Simard's work exemplifies Le Doeuff's concept of the imaginary. And, unusually, Simard herself provides us with the explicit narration of the work of the maternal imagery in the scientific enterprise. In her 2021 memoir, *Finding the Mother Tree*, Simard describes her intimate relationship with the forests and organisms that she studies, and narrates the context of discovery of the Mother Tree in a highly personalized way, appropriate to a memoir. Her musings on the relations between mature trees and seedlings began not in the lab but while on holiday, sitting under an old Douglas fir wondering how the seedlings on its outskirts could thrive even in very dry conditions when the seedlings planted in clear cuts (i.e. far away from any mature trees) were dying even with plenty of water.[35] Later, on a hike, Simard and a friend are forced to take refuge in two trees from a mother bear with cubs. Once safely back on ground she contemplates the relation between the older tree that is likely the parent of the younger one and which seemed 'protective of it, of all of us. I tipped my hat in thanks and whispered that I would be back to learn more from her' and seek 'insight into the mysterious manners of mothers in forests'.[36] When her research began to demonstrate the extent and importance of the interconnectedness of the forest organisms, Simard explains her growing appreciation of the fact in explicitly anthropomorphic terms: 'Ecosystems are so similar to human societies – they're built on relationships. The stronger these are, the more resilient the system.' She compares the ecosystem to an orchestra, to the human brain and to family relations: 'The cohesion of biodiversity in a forest, the musicians in an orchestra, the members of a family growing through conversation and feedback, through memories and learning from the past… They are complex. Self-organizing. They have the hallmarks of *intelligence*.'[37]

During this period of her research Simard gave birth to two children and her narrative links her experience of motherhood to her research. She explains how the idea of the Mother Tree, the name for the largest, oldest, most well-connected trees in a forest,[38] came to her immediately after thoughts of her own children (she had just thought about teaching them to make a poultice of baking soda for wasp stings):

> The old trees were the mothers of the forest.
> The hubs were *Mother Trees*.
> Well, mother *and* father trees, since each Douglas-fir tree has male pollen cones and female seed cones.
> But… it felt like mothering to me. With the elders tending to the young. Yes, that's it. *Mother Trees. Mother Trees* connect the forest.[39]

Like the plant advocacy literature that we discussed in Chapter 1, Simard's anthropomorphic *model* is deployed in the service of a *moral* anti-anthropocentrism:

> Our modern societies have made the assumption that trees don't have the same capacities as humans. They don't have nurturing instincts. They don't cure one another, don't administer care. But now we know that Mother Trees can truly nurture their offspring. Douglas firs, it turns out, recognize their kin and distinguish them from other

families and different species. They communicate and send carbon, the building block of life, not just to the mycorrhizas of their kin but to other members of the community. To help keep it whole. They appear to relate to their offspring as do mothers passing their best recipes to their daughters.[40]

In Peter Wohlleben's popular account of Simard's ideas this maternal imagery is extended even further. Simard, according to Wohlleben, discovered 'maternal instincts' in trees; passing nutrients through root systems '[y]ou might even say they are nursing their babies';[41] Mother Trees 'suckle their offspring'.[42]

Simard, of course, is perfectly well aware of the scientific taboo on anthropomorphism.[43] But her insistence on the avowedly anthropomorphic idea of the Mother Tree is part of a critique of the narrow, scientistic, short-termist and instrumental view of nature as a potential resource to be exploited. She does not say the word 'capitalism' but that is the word for it.[44] Simard associates this narrow view with a peculiarly modern malaise: disconnection from nature. To evoke the Mother Tree is both to signal that the scientistic, instrumental view is not her ways of doing things – the maternal botanical imaginary is a participatory element in the scientific enterprise – and that this was also not the way of the indigenous or First Nations peoples evicted from the lands on which her research now takes place. Further, as Simard acknowledges, some of what comes as a recent revelation to Western science is foundational to various indigenous schemes of thinking – that the forest (and beyond) is an interconnected whole; that there are underground root-fungal networks; that some trees, which some indigenous peoples have long called Mother Trees, have a particularly important role to play; and that an ecosystem is a matter of dynamic balance. Invoking principles common to indigenous peoples that express the idea that 'we are all related' or 'we are one', Simard is explicit that 'this kind of transformative thinking is what will save us'.[45]

To criticize the anthropomorphism of the idea of the Mother Tree seems churlish, but it is not beyond analysis. The form in which the Mother Tree appears in this popular science is of course – it can only be – rather superficially related to any indigenous version. It may resound with the faint echo of indigenous wisdom, but as we have it in Simard and Wohlleben, for example, it is an expression of a primarily Western scheme of thought. Because of this context there is reason to pause and think about the implications of the idea of the Mother Tree, even while we acknowledge the important role that it plays in Simard's work in bringing the climate catastrophe to attention and in seeking practical solutions (regenerating forests and keeping carbon in the ground). For there are no fathers in Simard's or Wohlleben's forests, in the sense that there is no paternal care. Nurture, care, upbringing and succour are exclusively maternal functions. The fact that the Mother Tree is in fact not exclusively female (the Douglas firs that are its inspiration are monoecious), or that 'maternal care' is an idea available for metaphorical extension, does not mitigate the immediate fact that the a priori association of care with the mother is itself a social problem that has long been the object of feminist criticism. Must it always be the mothers who 'look out for the whole neighbourhood'?[46] Is this idealized version of the mother the right image for thinking collective responsibility today?

Indigenous knowledge

In order to understand the nature of our objection to Simard's anthropomorphic model, we need first to think a little more about its potential relation to what is broadly known as 'indigenous wisdom', 'indigenous knowledge' or 'local knowledge'. UNESCO defines 'indigenous knowledge' as follows:

> Local and indigenous knowledge refers to the understandings, skills and philosophies developed by societies with long histories of interaction with their natural surroundings. For rural and indigenous peoples, local knowledge informs decision-making about fundamental aspects of day-to-day life.
>
> This knowledge is integral to a cultural complex that also encompasses language, systems of classification, resource use practices, social interactions, ritual and spirituality.[47]

Indigenous knowledge in this sense encompasses both knowledge content and ways of knowing, although the two aspects are most often thought as inextricable and co-constituting. Although specific aspects of indigenous knowledge (especially skills and determined, local knowledge claims) may easily be incorporated into Western science, for many the Western scientific method is fundamentally at odds with indigenous ways of knowing. Institutions (especially educational institutions) of Western science in settler colonies are increasingly aware of the need to incorporate indigenous knowledge into curricula and field work. Simard's research and teaching is certainly sensitive to this. But for many on both sides Western science and indigenous knowledges remain distinct and perhaps even incompatible.

The problem of negotiating the relationship between Western science and indigenous knowledge is one of the main themes of Robin Wall Kimmerer's *Braiding Sweetgrass* (2013). As a member of the Citizen Potawatomi Nation and a scientist Wall Kimmerer was early painfully aware that '[t]o walk the science path I had stepped off the path of indigenous knowledge' and her work bears witness to her attempts to address this and to become a 'traveller' between the two.[48] Wall Kimmerer distinguishes between the practice of science (which can work with indigenous knowledge) and the scientific world view (which cannot). In the scientific worldview 'a culture uses the process of interpreting science in a cultural context that uses science and technology to reinforce reductionist, materialist, economic and political agendas', based on 'the illusion of dominance and control [and] the separation of knowledge from responsibility'. Wall Kimmerer's 'dream' (which is also in large part the reality of her scientific work) is of 'a world guided by a lens of stories rooted in the revelations of science and framed with an indigenous worldview'.[49]

Throughout *Braiding Sweetgrass* Wall Kimmerer relates some of the stories and ceremonies associated with the Anishinaabe First Nations (including the Potawatomi Nation) associated with the Great Lakes regions of what is now Canada and the United States. But, she asks, what meaning can these stories have for us today?[50] The story of her sincere and steadfast but nevertheless faltering efforts to learn the language of her Potawatomi tribe is emblematic of the difficulty inherent in this question. Wall Kimmerer is clear that she will never be a fluent speaker, indeed will probably not pass beyond the very basic. But she is able to understand and to communicate something of the specificity of the language in contrast with her native English. In the Potawatomi language, she tells us, both nouns and verbs can be either animate or inanimate, depending on whether their referent is alive, and all parts of the natural world (including the landscape, water ways and so on) are alive and are met as commanding respect as co-residents of the natural world.[51] Becoming 'bilingual between the lexicon of science and the grammar of animacy' then means to practice science 'with awe and humility', with a sense of responsibility towards the 'more-than-human world'.[52]

This worldview is expressed in the teachings known as the Original Instructions of the indigenous peoples of the Great Lakes area, which include stories and ceremonial practices. Wall Kimmerer sees the 'withering away' of ceremony in the 'dominant society' as a symptom of its alienation from nature and from forms of collective self-understanding.[53] But again, she asks, what meaning can indigenous ceremony – often focused on other species or on the seasons, for example – have for us today? The Western appropriation of such First Nations ceremonies is both morally reprehensible and somewhat absurd (or at the least shallow and artificial). So how do we learn from them? Wall Kimmerer offers no easy solutions to these problems but sees value in the 'organic' creation of ceremony and tradition in the spirit of the Original Instructions, which allows her to practice science in the spirit of the Original Instructions.[54] Where science aims to understand, respond to and mitigate climate change (including via ecological restoration), it is not hard to see that an appreciation of the complex interrelations and interdependencies of living beings is a clear scientific advantage, not a moral posture.[55] But Wall Kimmerer is no Polly Anna about this. She, as a First Nation citizen, remains acutely aware of her distance from the teachings of the Potawatomi elders, unable to fully inhabit the world created and indeed sung by the Potawatomi language. Wall Kimmerer's book is thus also a warning to Western readers, as approaching indigenous knowledge means understanding the extent to which you do not understand.

The difficult relationship between Western science and indigenous knowledge(s) is in some respects mirrored in the difficult relationship between Western philosophy and indigenous philosophies. Dale Turner's reflections on this in the context of his work on Aboriginal rights in Canada and indigenous intellectual culture are helpful to understanding this. Turner, a political philosopher and a member of the Temagami First Nation (which is also one of the Anishinaabe Nations), identifies three indigenous projects in philosophy. The first he calls 'doing "indigenous philosophy" proper. This kind of activity involves highly specialized forms of thinking; it is a distinctively indigenous activity'.[56] The second is doing European philosophy as an indigenous person, promoting cross-cultural philosophical dialogue with the aim – importantly – of making a political difference to indigenous lives.

The third – his own kind of philosophy – is the *critical* engagement of European ideas as a philosophical exercise and a political activity, making 'an investigation of the meaning and praxis of colonialism a central activity of an indigenous intellectual community'.[57] Turner is clear that the first project, to the extent that it might happen within non-indigenous contexts and with the aim of educating non-indigenous persons, is fraught with inherent epistemological problems because such philosophy is rooted in oral tradition and in indigenous languages. It remains to be seen, he writes, whether indigenous philosophies of knowledge 'can be articulated in English'. Although 'finding the right "place" for terms like "spirituality" is essential to a critical indigenous philosophy', Turner refuses to put himself in the 'privileged position' of explaining indigenous spirituality because, he says, he himself is not an indigenous philosopher 'proper'.[58] Once again, then, Turner's book can be read as a warning to the Western reader. Do not excuse yourself (or anyone else) from listening to indigenous philosophers and intellectuals and understand the urgency of 'asserting and protecting the rights, sovereignty, and nationhood of communities',[59] but do not think that indigenous philosophy is necessarily compatible with Western philosophy and do not think that elements of it are just going to slot right into the Western curriculum or idiom.

When discussing indigenous philosophies from within a Western standpoint, that standpoint can of course not simply be shed in favour of another. But developments within anthropology in the last decades do at least help us to understand the historical specificity of that standpoint, or the historical and cultural specificity of the basic form of Western metaphysics, which is too quickly assumed to be universal. When we can at least think of comparing the deeply embedded 'schemas', as Philippe Descola calls them, of Western and indigenous ontologies, the mark of necessity that can seem to hang over the former begins to fade.[60] But more than this, an anthropology that aspires to be a philosophy can signal a way out of the kind of impasse that, in similar ways, both Wall Kimmerer and Turner butt up against in trying to think the relation between indigenous knowledge and science or between indigenous and Western philosophies, respectively. This is what we see in the work of Eduardo Viveiros de Castro, where we find another way of thinking about anthropomorphism, and thus another way of asking what kind of anthropomorphism is at issue in Simard's idea of the Mother Tree.

'When everything is human, the human becomes a wholly other thing'

On the basis of anthropological fieldwork with the Arawaté people of northern Brazil, Viveiros de Castro has developed the concepts of what he calls Amerindian 'perspectivism' and 'multinaturalism'. He positions himself against many of the presuppositions of traditional and 'postcolonial' anthropology, and chiefly amongst these against the idea that Western anthropology is condemned only to 'represent' or 'invent' the 'Other' (the 'object' of the anthropologist's study) according to the interests and established concepts of the West. This idea, which might seem to be an admirable admission on the part of anthropology, is for Viveiros de Castro a narcissistic 'complacent paternalism' that makes even 'a certain theoretical postcolonialism the ultimate stage of ethnocentrism'[61] because it effectively denies the so-called others a 'speaking part' and condemns them always to be only the object of the anthropologist-subject. Instead, for Viveiros de Castro 'every nontrivial anthropological theory is a *version* of an indigenous practice of knowledge'.[62] What he calls the Amerindian concepts of perspectivism and multinaturalism are therefore not the 'authentic' elements of an indigenous worldview, of its 'cosmological beliefs, unconscious schemas' or 'mental categories' but an outcome of the 'relational synergy between the conceptions and practices of the worlds of [anthropology's] "subject" and "object"'.[63] In Viveiros de Castro's case, this is the synergy between Amerindian cosmology and a certain Western philosophy.[64] What he proposes is not an interpretation of Amerindian thought but 'an experimentation with it', and thus also with his own Western thought. To take indigenous thought seriously is not to explain, contextualize or rationalize it (not, for example, to make is assimilable to Western science) but to draw out its consequences for one's own thought.[65] It is not a matter of whether the Western thinker can or does 'believe' what the Arawaté people, for example, think, not least because the Arawaté system of thought is itself not a set of 'beliefs' (just as we do not characterize Western philosophy as a set of beliefs). Quoting Godfrey Lienhardt, 'it is not finally some mysterious "primitive philosophy" that we are exploring, but the further potentialities of our thought and language'.[66] Thus indigenous thought is relegated neither to a kind of common, inaccessible alterity nor to the sentimental '"traditional knowledges" so lusted after in the global market of representations': 'neither derealizing [it] as fantasies of the other nor fantasizing that [it] is actual for us'. We Westerners cannot think *like* the Arawaté, Viveiros de Castro writes, but we can think *with* them.[67]

The concepts of perspectivism and multinaturalism are the outcome of such a thinking *with*. They do not describe an authentic indigenous way of thinking that is either absolutely other than or superior to Western philosophy; they are a possibility for a Western philosophy that is not just business as usual. They are the outcome of a concerted effort to think otherwise with unfamiliar thoughts. In the briefest summary, perspectivism is the indigenous theory that the way that humans perceive animals, gods, the dead and all other 'subjectivities that inhabit the world' is profoundly different from the way that these beings see humans and other subjectivities.[68] All such subjectivities have a 'soul', see themselves as persons, and therefore are persons. But they do not, under normal circumstances, see other kinds of beings as persons. Sharing the same kind of soul, all subjectivities see themselves as humans, while seeing all other kinds of beings as non-human:

> Typically... humans will, under normal conditions, see humans as humans and animals as animals (in the case of spirits, seeing these normally invisible beings is a sure indication that conditions are not normal: sickness, trance and other 'altered states'). Predatory animals and spirits, for their parts, see humans as prey, while prey see humans as spirits or predators... In seeing *us* as nonhumans, animals and spirits regard themselves (their own species) as human: they perceive themselves as (or become) anthropomorphic beings when they are in their houses or villages, and apprehend their behaviour and characteristics through a cultural form: they perceive their food as human food – jaguars see blood as manioc beer, vultures see worms in rotten meat as grilled fish – their corporeal attributes (coats, feathers, claws, beaks) as finery or cultural instruments, and they even organise their social systems in the same way as human institutions, with chiefs, shamans, exogamous moieties and rituals.[69]

According to this perspectivism what we humans share with all animals is thus not animality but humanity, because each species is human to itself, although none are (normally) human to each other.[70] This perspectivist theory is not just another version of Western relativism, because it does not say that there is a plurality of points of view each of which is just as true as the other; rather, 'where humans are concerned [for example], there *is* a true and accurate representation of the world', and it is crucial to keep perspectives separate from each other, because it is important that a human does not see worms in a cadaver as grilled fish or that they do not mistake a jaguar for their friend. But anyway, perspectivism does not postulate a plurality of different representations of the same world, but a common representation of different worlds: what *is* blood for us *is* beer for jaguars. Thus the quasi-epistemological notion of perspectivism passes into the 'veritable ontological one of multinaturalism'.[71] In this epistemo-ontological regime the 'functions' attributed to soul and body are the inverse of the Western attribution. All beings share the same kind of soul and thus 'are human' and represent things in the same way, but different kinds of beings have different kinds of bodies and the body (not a bare physiological phenomenon but its strengths and weaknesses, what it eats, how it moves, where it lives and so on) is the point of view.[72] Viveiros de Castro connects this with the mythic discourse of a pre-cosmological condition in which 'the corporeal and spiritual dimensions of beings do not yet conceal each other'; that is, all beings are and see each other as human. But there is not thereby a primordial identity between humans and non-humans; rather, the feline and the human dimensions of the jaguar, for example, are not yet bifurcated in the cosmological process. Cosmological myths (which for Viveiros de Castro have the same intellectual status as pre-Socratic Western cosmologies) narrate the sorting of the pre-cosmological differences into mundane modes of existent where each is human to itself but none are human to each other. But the jaguar, for example, having formerly been human, continues to be so 'even if in a way scarcely obvious to us'.[73]

If here the word 'human' seems to have an exceptionally equivocal function, that is exactly the point: 'What perspectivism affirms, when all is said and done, is not so much that animals are at bottom like humans but the idea that as humans, they are at bottom something else'.[74] (Or, as Ursula le Guin's character Harfex says: 'Call them trees, certainly... They really are the same thing, only, of course, altogether different'.[75]) This is an 'anthropomorphic presupposition', because it assumes that 'the primordial assumes a human form'. But this anthropomorphism is 'radically opposed to the persistent anthropocentric effort in Western philosophies', because – and this is the most important point for our purposes – 'When everything is human, the human becomes a wholly other thing'.[76] No doubt it is not easy to think this through, but that again is the point. We will not 'expand the still excessively ethnocentric horizons of 'our' philosophy'[77] by staying within familiar conceptual terrain.

Bearing this in mind, what kind of anthropocentrism is at work in Simard's idea of the Mother Tree, and in its reception in the popular scientific literature? What happens when we compare it to the anthropocentrism of Viveiros de Castro's Amerindian perspectivism? And what does all this have to do with plant philosophy and vegetal sex?

Two anthropomorphisms

The difference between the anthropomorphism in Viverios de Castro's account of Amerindian perspectivism and the anthropomorphism that is the mainstay of much of the plant advocacy literature (let us call this 'traditional anthropomorphism') is that whereas the latter has recourse to what is most familiar and indeed comforting to us in order to make plants seem like us, the former (perspectivist anthropomorphism) defamiliarizes and discomforts. The traditional form of anthropomorphism projects a conception of the human onto animals and plants without the meaning of 'the human' or of any human traits and phenomena being put into question. The more challenging perspectivist anthropomorphism, on the other hand, has as its basis the equivocation over 'the human'. ('When everything is human, the human becomes a wholly other thing.') When Simard and Wohlleben speak of trees 'mothering' saplings and ascribe nurturing or maternal instincts to them, this is an interpretation of a set of empirical facts via an uncritically accepted, idealized image of human mothers defined by their responsibility for the functions or duties of care and nurture. To that extent it reinforces that image, not just as it extends to plants but even more so as concerns its human 'source'. Wall Kimmerer evokes something like the Amerindian perspectivism laid out by Viveiros de Castro when she writes that 'In the old times, our elders say, the trees talked to each other. They'd stand in their own council and craft a plan'. Western science dismisses this, she writes, because 'the conclusion was drawn that plants cannot communicate because they lack the mechanisms that *animals* use to speak. The potentials for plants were seen purely through the lens of animal capacity'. This kind of anthropomorphism, which she associates (in contrast with the animist grammar of the Potawatomi language) with the arrogance of English assumes that 'the only way to be animate, to be worthy of respect and moral concern, is to be a human'.[78] But when a moral anti-anthropocentrism like that of Wall Kimmerer, Simard or Wohlleben goes no further than to ascribe human attributes and kin relations to plants *without reflecting back on the meaning of those attributes and kin relations*, how does this avoid the animal lens? How does it avoid the unidirectional anthropomorphism that always and only sees humanity *in its most familiar and unquestioned form* as the source of significant modes of existence? The problem is two-fold when it comes to Mother Trees. Are there not ways that trees might co-exist, be implicated in each other's life and growth, even co-live and co-grow with their own and other species, that are not simply reflections of human social forms? And are there not ways of thinking nurture and care in human existence that do not immediately evoke an idealized version of the maternal? Is it possible that the ways of existence of trees might provide alternative conceptual forms for thinking about care that relieves human mothers of their (imagined) exclusive responsibility for care? Would that not be the way to think plants beyond the animal lens?

As we have already seen in Chapter 1 the philosophical botanist Jacques Tassin objects strongly to the traditional anthropomorphism that, in his view, disfigures discussion of plant life. Tassin characterizes this anthropocentrism as a kind of magical thinking which sees in plants only a reflection of ourselves. This results in 'phantasmagorias about trees', including that they live according to (human) familial patterns. But, Tassin writes, 'Nature does not ask trees to function like humans'. The only way to understand plant life is in relation to its 'full alterity': 'To rediscover the tree is first of all to rediscover alterity – not a projection of ourselves but an alterity that we accept has an unknown and inaccessible part.'[79] But does Tassin's insistence upon absolute alterity question the source terms of traditional anthropocentrism anymore that that anthropocentrism itself? Is the invocation of the in-principle unknown and inaccessible not akin to the anthropological narcissism that Viveiros de Castro identifies – the idea that we can only ever see ourselves in plants, and, moreover, as we are already familiar to ourselves?[80] Is it possible that, beyond the choice between traditional anthropomorphism and absolute alterity, something of the structure of perspectivist anthropomorphism might open a new path for plant philosophy?

One of the most productive aspects of Viveiros de Castro's *Cannibal Metaphysics* is that the forced reconceptualization of the human – even if, for the moment, we get no further than the first step in this process, no further than an intellectual acceptance of the possibility of such a reconceptualization – is emblematic of a far more extensive project of conceptual reimagination. Each time a familiar term is introduced, its place in the anthropological and philosophical context of 'thinking with' transforms it, or at least demands that it be transformed. This is true of even the most basic methodological concepts of anthropology, including that of 'comparison'. Comparison is not the placing side by side of two or more sociocultural entities external to the observer: 'Comparison, as I conceive it, on the contrary, is a "constitutive rule" of method, the procedure involved when the practical and discursive concepts of the observed are translated into the terms of the observer's conceptual apparatus.'[81] The observer is of course implicated in this translation. Often that is taken to mean that the observer inevitably imposes something of themselves on the translation, and no doubt that is always one aspect of things, especially if one translates in order to 'explain, generalise, interpret, contextualize' and so on. But in the construction of anthropological theory 'translation' is the end, not a means. The theory *is* a translation. Good translation 'succeeds at allowing foreign concepts to deform and subvert the conceptual apparatus of the translator such that the *intentio* of the original language can be expressed through and thus transform that of the destination.'[82] Anthropological theory as a translation (of, in Viveiros de Castro's case, Amerindian concepts) may thus be not a transposition of one set of source concepts into another set of destination concepts but the transformation of the destination. At its most ambitious in Viveiros de Castro's case this means the transformation of Western philosophy.[83]

What is more, perspectivism is itself an 'image' of translation, which is the image of a 'controlled equivocation, of referential alterity between homonymous concepts', most notably, of course, the concept of the human. Equivocation is thus not a communicative pathology but 'a properly transcendental category, a constitutive dimension of the project of cultural translation proper to the discipline... To translate is to take up residence in the space of equivocation.'[84] Accordingly, incommensurability is no impediment to comparison; indeed, Viveiros de Castro writes, 'only the incommensurate is worth comparing.'[85] This is the sense in which something of the structure of perspectivist anthropomorphism might open a new path for plant philosophy, beyond the choice between traditional anthropomorphism (as in most of the plant advocacy literature) and the blank inaccessibility of the absolute alterity of plants.

What happens, with this model of comparison-translation-equivocation, if we compare animals and plants?

Of course, there already exists a tradition of animal-plant comparisons. In previous chapters we have shown how this tradition proceeds, for the most part, via analogies between animals and plants, where the priority and the dominance of the animal model are most often not contested, and thus that and how, in the case of sex and the terminology of sexual reproduction, analogy becomes identity. More recently plant philosophy has renewed the idea of the comparative tradition but with the explicit aim of moving beyond the zoological model to articulate the specificity of plant life. Francis Hallé undertakes his comparison from a primarily biological context. To take just one example from his *Éloge de la plante*, the comparison of the evolutionary 'strategies' of animals and plants suggests very convincingly that many of the shibboleths of evolutionary theory are inadequate for understanding plant evolution (i.e. not the history of plant evolution, but the mode of evolution of plants).[86] The philosopher Florence Burgat's use of a comparative method (as a starting point, at least) is explicitly proposed as an alternative to the dead ends of traditional anthropomorphism and a certain interpretation of the consequences of the radical alterity of plants: the reduction of plants to physio-chemical ensembles without being able to say anything about them as *living* beings.[87] Burgat's approach acknowledges the limitation that drives both these positions: we cannot put ourselves in the place of a plant, or we cannot replace anthropocentrism with phytocentrism. But an initially negative comparison between animals and plants, according to which we can at least say what plants are not, is necessary if we are to begin to discern the outlines of the radical alterity of vegetal vitality.[88] This is because such a comparison shows us, according to Burgat, that there is nothing in common between vegetal and animal life, hardly even 'life' itself, as what life *is*, from Burgat's phenomenological point of view, is so radically different for plants and animals.[89]

However, for both Hallé and Burgat the result of the comparison, which separates plants and animals, leaves the conceptual terrain of the abandoned animal model intact. According to Hallé, 'even those fundamental concepts such as the individual, the genome, sexuality, species or evolution need to be developed, or transformed, if we are to achieve an objective biology, rid of zoocentrism and anthropocentrism'. But this does not seem to mean that we need new *generic* biological concepts, but rather additional ones adequate to plant life, as we currently only speak an 'animal language' which does not allow us a relation to 'vegetal truth'.[90]

But what if we tried to take this one step further? What if we attempt the step from the comparison that separates to the translation that transforms? What if, with everything that we have learned about the trouble with vegetal sex, we tried to understand what the comparison between animal and plant sex means for the generic concept of sex and its terms 'male' and 'female'? What if, when both animals and plants are sexed, 'sex' becomes a wholly other thing?

EPILOGUE: VEGETAL SEXUALITY AND US

Throughout this book we have tried to show, against the received histories of botany, that there is nothing obvious about plant sex. We have argued, further, that the specificity and complexity of vegetal sex emerge in the history of botany where botany is at its most philosophical or is interpreted philosophically. We have shown that it is not obvious what 'male' and 'female' mean in relation to plants, and indeed that this problem has been an issue in both plant philosophy and in some modern botany – for example in Charles Darwin's work on flower forms, in the terminological debates of the late nineteenth and early twentieth centuries and more recently in David Lloyd's oeuvre. We have also argued that the treatment of the problem in the plant philosophy tradition does not take the meaning of 'male' and 'female' for granted and simply ask whether and how these categories apply to plants. Rather, in properly philosophical fashion, it asks, 'What are "male" and "female"?' And, as we have seen, despite the common-sensical assumption that we all know what male and female are, the question is surprisingly hard to answer in relation to plants.

What, then, does this mean for the generic concepts of male and female? What does this mean for the generic concept of sex? If, as we have argued, plant sex cannot be adequately thought according to the animal model, and if the generic concepts tend to assume the animal model, how can we continue to use 'sex' generically, to refer to a biological phenomenon common to almost all forms of eukaryotic life? What is the nature of that commonality – the commonality of male and female – across the realms of animal, vegetable, fungi and protista?[1] And what might the identification of this commonality mean for the popular, untechnical use of the notions of sex, male and female?

The question of the commonality of male and female across different realms of life – and here, specifically, their commonality across plant and animal life – can be cast in terms of the biological distinction between homology and analogy. Biologists have for a long time used these terms to distinguish different kinds of similarity when comparing structures and parts in different animals. Richard Owen is credited with having first defined these terms in something like their modern biological senses, in the glossary to his 1843 *Lectures on Comparative Anatomy and Physiology of the Invertebrate Animals*: 'Analogue. A part or organ in one animal which has the same function as another part or organ in a different animal... Homologue. The same organ in different animals under every variety of form and function'.[2] According to these definitions the wings of insects and birds are analogues (they are different kinds of parts, though they have the same function), whereas the radius and ulna bones in the human and the bat are homologues (the same organ, albeit they may be thought as performing different functions). Similarly, human male and female sex organs can be said to be homologous since they develop from the same embryonic tissue. Since the widespread acceptance of the theory of evolution, homology and analogy are defined in terms of descent: homologues are similar due to common descent; analogues are similar despite no common descent and are thus the result of convergent evolution, that is, the independent evolution of similar traits in organisms with no common ancestor with that trait. Homologous traits (in, for example, human beings and lizards) are inherited from a common ancestor which also had that trait. Biologists today also speak of genetic homologies across species, and part of the process of meiosis is described in terms of the pairing of homologous sets of chromosomes from each parent.[3]

Are there any homologies between animals and plants, at any of these levels? There are certainly genetic homologies. For example, there is an homologous region of non-coding DNA which can be presumed to have been inherited from a common ancestor – indeed the Last Eukaryotic Common Ancestor (LECA), as it is known (though *what* it was is not known).[4] This common ancestor is thought to have been sexual (to exhibit sexuality), but this does not necessarily mean that there was a distinction of sex (male and female); indeed, many think that there probably was not.[5] It is necessary, then, to distinguish between 'sex' or 'sexuality' and 'sexes' (male and female). This gives us a terminological problem, because 'sex' is used to refer to both the process of sexuality and the nature of the difference between male and female. For the rest of this discussion, though, we will use 'sexuality' for the former' and 'sex' for the latter.

Sexuality is a genetic mixing process in principle distinct from reproduction. There are two main (and competing) definitions. According to one definition, sexual mixing is any genetic exchange between individuals, which results in a genetically new (i.e. changed) individual but not necessarily an increase in the number of individuals. According to this definition, perhaps most closely associated with Lynn Margulis, simple forms of sexuality occur amongst viruses and bacteria. In bacterial conjugations, for example, one donor bacterial organism 'donates' genetic material to a 'recipient' bacterial organism which incorporates this material (the process of genetic recombination).[6] According to the second definition, sexual genetic recombination must include the process of meiosis, which excludes genetic recombination in viruses and bacteria from the definition of sexuality.[7]

For both definitions, sexuality does not necessarily require sexes. Indeed, as Beukeboom and Perrin say, sexes, male and female – defined as the functions producing oogametes or the oogametes themselves – 'are latecomers in the long history of sex'.[8] Many organisms have, instead, what are called 'mating types' which may be distinguished from sexes either in being isogamous or in being multiple (as we saw in Chapter 5 with Darwin's heterostyled flowers). Isogamous dual mating types are often marked as '+' and '–', but mating types may number in the hundreds in fungal species.[9]

Again, according to both definitions, we could say that the *process* of sexuality is ancestrally homologous in plants and animals. But it is not clear that there is any meaningful sense in which there is any homology between animal and plant *sexes*, whether homology is thought at the genetic, evolutionary or morphological level. The point, indeed, is more general than the contrast between animals and plants: 'Certain cellular aspects of sexuality are nearly identical. Nonetheless, many, indeed the vast majority of sexual manifestations that conspire to bring together mates are analogous rather than evolutionarily homologous. That is, similar sexual systems are not usually related through descent but evolved independently.'[10] The distinction between male and female (anisogamy) is thus an evolved species trait. It is believed that this distinction has evolved independently several times; that is, not all anisogamous species trace the anisogamy trait back to a common anisogamous ancestor.[11]

The issue is as much conceptual as it is one subject to empirical or speculative-scientific investigation, not least because there exist different ways of thinking about biological homology. But in the broadest terms it is surely right to say that whereas specific male and female forms (e.g. in various mammalian species) may be homologous, the generic concepts of male and female, and thus of sex, in their greatest generality, can only refer to relations of analogy, and that the comparison between male and female in plants and animals is thus, fundamentally, an analogy. This also means that the generic concepts of 'male', 'female' and 'sex' are themselves fundamentally analogous terms; that is, at the highest level of generality they can *only* refer to an analogous having in common.

Thus the early plant philosophy tradition was right: there are only analogies between male and female animals and plants, especially at the level of the individual and the part or organ. *Strictly speaking* there is also only an analogy at the level of the gamete. Indeed, this is the *only* meaningful analogy between plant and animal sex; there is no meaningful biological analogy at the level of the organ or the individual, only a crude metaphorical transfer from animality to vegetality.

What, then, is the specificity of vegetal sex? This question can only be answered by sloughing off the inappropriate anthropomorphic ways in which it has hitherto been thought. Plants do not live in gender-segregated societies in which individuals find themselves subject to normative expectations concerning their social roles, behaviour and physical being. Plants do not live in or have the responsibilities of nuclear or even blended or chosen families or other social groups. They cannot be gay or straight, cis or trans, chaste, abstinent or promiscuous. From one perspective this is their peculiar and enviable ontological privilege, especially in comparison with the burden of normative gender borne by human beings: nothing is normal, because nothing is abnormal. Vegetal sex implies no duties, no right or wrong. It has no moral or social implications at all. Vegetal sex just is, without the burden of any ought.

The specificity of vegetal sex, then, is that it is entirely reduced to sexuality, in the sense defined above. Vegetal sexuality is *in principle* able to be thought entirely independently of any ideological interpretation of its fact. Arguably, this is what some plant scientists have implicitly tried to demonstrate in being rigorous about the use of a specifically vegetal terminology, as far as this is possible. This is so even if we find it appropriate (as all plants scientists do) to name vegetal anisogametes with the terminology of male and female because, as Coulter put it: 'it is useless to ask what "maleness" and "femaleness" are, but it seems evident that they represent essentially conditions of the nucleus'. We call heterogametes 'male' and 'female', but these are 'words rather than explanations', standing in for 'a definite physiological difference'.[12] No moral or social duty and no subjective identity can possibly follow on from this for the plant; it makes no sense at all to attach a social expectation or subjective identity to the bare biological fact of the production of gametes, or indeed to their absence.

But of course, if this independence of vegetal sexuality from ideology is so in principle, it is not so in most historical practice, as the extensive anthropomorphic discourse of plant sex from Vaillant and Linnaeus to today's popular scientific literature shows. The primary ruse in the anthropomorphizing of vegetal sex is the conflation of anisogamous vegetal sexuality with sexed organs or individuals – the anthropomorphic personification of the flower or of the plant – a conflation facilitated by the use of the same words (sex, male, female) in relation to each. The philosophical impulse of the plant philosophy tradition was to ask, 'What do we mean when we speak of "male" and "female" in relation to plants?' It is when this kind of philosophical question is dismissed as irrelevant to botany that the path to anthropomorphism is cleared. But it is also, as we have tried to show, when the question nevertheless gets asked, by the philosopher in the botanist, that we see explicit attempts to avoid anthropomorphism, or that we see at least that the question of the specificity of vegetal sex arises.

It is striking that the major ideological move in the thinking of human sex adopts a similar method to the anthropomorphizing of plants: the conflation of meaning at the levels of the gamete, the organ and the individual. It is again ambiguities arising from a shared, imprecise terminology that facilitates this. As we indicated in Chapter 4, it is particularly difficult, when the topic is sex, to separate out the strictly biological meaning of sexuality and sexes from their numerous ideological and otherwise socially inflected meanings. That is, it is extremely difficult to separate out what is the case from the social and cultural presuppositions of what ought to be. This, indeed, is what the anthropomorphizing of vegetal sexuality really shows – nothing at all about plants but quite a lot about humans and their dominant gender and sexual ideologies, ideologies whose claim to universality demands that we see the same things in plants that we see in human social arrangements. That is one of the reasons why the history of analogies between animals and plants always moves from the former (as source or model) to the latter (as ideological sink or copy).

But what if the analogy could go the other way for once?

If, as we have suggested, the only meaningful analogy between vegetal and animal sex is at the level of the gamete (the animal gamete also being the only functional, living part of the organism that buds off or out vegetatively), we could say that the commonality of sex or the only meaningful having-in-common of male and female across kingdoms Animalia and Plantae – an essentially analogical commonality – is a shared vegetal sex.[13] At this level, human vegetal sex (sexuality [genetic recombination] and anisogamy) also just is, without the burden of any ought, and also can not *in itself* form the basis for any subjective or social identity. Human vegetal sex (sexuality and anisogamy) at this level is the identification of a species trait which belongs precisely to the abstraction 'the species' (and not necessarily to every individual within the species) just as it belongs to all angiosperm species, for example. (Vegetal) sex as a species trait is conceptually quite distinct from any discourse of the individual, because any possible talk about human individuals and their characteristics or about groups of individuals is value-laden talk in which everything that is foreign to vegetal sex is determining: history, psycho-sociality, morality, culture, ideology and metaphysics (remembering that the 'individual' is a metaphysical category).[14]

It takes some effort to maintain this conceptual distinction because the historical tendency to project what belongs to the realm of human psychosociality on to vegetal sex has seemed almost irresistible – as the history of discussions of the sex of plants has shown. However, recognizing the force of this tendency and attempting to be vigilant against it (where conceptual discrimination, or philosophy, is one form of vigilance) is at least to recognize the extent to which the determination of our understanding of vegetal sex has been inevitably social and historical. The basic ideological move, on the other hand, is to reverse this, imagining that vegetal sex – which is the basic form of sex generically understood – determines society and history and individuals' places, roles and subjective identities within it. But 'vegetal sex' is not the name for the ontological or natural foundation of human psychosocial existence or subjective identities, not least because this existence and these identities can and do play out in the absence of gamete production and also, more importantly, in ways that render vegetal sex irrelevant. What would it mean for us to accept this understanding of vegetal sex as our own? What would it mean for us to accept the autonomy, in the last instance, of the historical, the psychosocial, the political and the ethical – the real foundation of all of our individual identities – from any determination by our vegetal sex, and vice versa? This would not be to deny the biological reality of vegetal sex as a human species trait but to affirm the psychosocial and historical determination of our sexed and/or gendered identities and to accept their open futures.

NOTES

Introduction

1   Michael Marder, *Plant Thinking: A Philosophy of Vegetal Life*, Columbia University Press, New York, 2013, 94.

2   Quentin Hiernaux and Benoît Timmermans, eds., 'Introduction', in *Philosophie du végétal*, Vrin, Paris, 2019, 8.

3   Hiernaux, 'Pourquoi et comment philosopher sur le végétal?', in *Philosophie du végétal*, 13.

4   Marder, *Plant Thinking*, 11. Later he writes that 'plants quietly subvert classical philosophical hierarchies and afford us a glimpse into a lived (and growing) destruction of Western metaphysics' (53).

5   Ibid., 18.

6   There is a feminist literature on the gendering of plants as female and on the gender politics of talk of plant sexuality, notably as it appears in the works of Carl Linnaeus and Erasmus Darwin. See, for example, Londa Schiebinger, 'Gender and Natural History', in N. Jardine, J. A. Secord and E. C. Spary, eds., *Cultures of Natural History*, Cambridge University Press, Cambridge, 1996; Patricia Fara, *Sex, Botany and Empire: The Story of Carl Linnaeus and Joseph Banks*, Icon Books, Duxford, Cambridge, 2003; Lincoln Taiz and Lee Taiz, *Flora Unveiled: The Discovery and Denial of Sex in Plants*, Oxford University Press, Oxford, 2017. This literature demonstrates the gendering of botanical practice and language and often concentrates on the metaphorical extensions characteristic of discussions of plant sex and on the gendered social contexts of these discussions. This literature is certainly relevant to plant philosophy, but it deals with different questions than those raised in this book.

7   Francis Hallé, *Éloge de la plante: Pour une nouvelle biologie*, Seuil, Paris, 1999, 328. Although Hallé includes 'sexuality' (along with 'individual', 'genome', 'species' and 'evolution') amongst those fundamental concepts of biology that need to be transformed in order to do justice to plant being (*Éloge de la plante*, 322), the topic is not tackled head-on in his work.

8   This distinction echoes Gaston Bachelard's distinction between the object of science and the object of the history of the sciences. See Bachelard, 'L'objet de l'histoire des sciences', in *Études d'histoire et de philosophie des sciences*, Vrin, Paris, 1970 [1968].

9   See, for example, James D. Mauseth, *Botany: An Introduction to Plant Biology*, Fifth edition, Jones & Bartlett, Burlington, MA, 2014, 2–3. Popular scientific genres are often less fastidious; for example, David Attenborough's 1995 television series and its accompanying book (*The Private Life of Plants: A Natural History of Plant Behaviour*, BBC Books, London, 1995) include plants, algae and fungi under the category 'plants'.

10 Taiz and Taiz, *Flora Unveiled*, vii.

11 The use of the term 'indigenous' is not unproblematic. As Linda Tuhiwai Smith points out (*Decolonizing Methodologies: Research and Indigenous Peoples*, Zed Books, London, 2012 [1999], 6), it risks lumping diverse histories and peoples into a generic group that does justice to none. However, it seems to be the best term to refer to the thought, knowledge and traditional skills developed by societies with ancestral claims to land and with those societies' relations to the natural world. Tuhiwai Smith uses the term because it 'internationalizes the experiences, the issues and the struggles of some of the world's colonized peoples... [it] has enabled the collective voices of colonized people to be expressed strategically in the international arena' (7). In this book, then, we follow Tuhiwai Smith's lead.

Chapter 1

1 Or even, as plant scientist Jack C. Schultz is quoted as saying in a BBC online article: 'Plants... "are just very slow animals". http://www.bbc.com/ earth/story/20170109-plants-can-see-hear-and-smell-and-respond (last accessed 14 January 2021).

2 Anthony Trewavas, 'What Is Plant Behaviour?', *Plant, Cell and Environment*, Vol. 32 (2009): 606.

3 For a definition of plant behaviour see, for example, Richard Karban, 'Plant Behaviour and Communication', *Ecology Letters*, Vol. 11 (2008): 727.

4 See, for example, Inyup Paik and Enamul Huq, 'Plant Photoreceptors: Multi-functional Sensory Proteins and Their Signalling Networks', *Seminars in Cell & Developmental Biology*, Vol. 92 (2019): 114–21; Martina Legris et al., 'Phytochrome B Integrates Light and Temperature Signals in *Arabidopsis*', *Science*, Vol. 354, No. 6314 (18 November 2016): 897–900. Neither of these papers claim that plants can 'see'.

The scientific literature on plant behaviour is obviously vast. The research referred to in this chapter has been selected either because of its acknowledged importance and influence or because it is the research that is referred to in the popular scientific literature.

5 María A. Crepy and Jorge J.Casal, 'Photoreceptor-mediated Kin Recognition in Plants', *New Phytologist*, Vol. 205 (2015): 329–38. The authors of this paper do not claim that the plants 'see' their kin.

6 Ernesto Gianoli and Fernando Carrasco-Urra, 'Leaf Mimicry in a Climbing Plant Protects against Herbivory', *Current Biology*, Vol. 24 (5 May 2014): 984–7. In leaf mimicry the leaves of climbing plants change in various ways (including size, shape and colour) to mimic those of the supporting tree. Gianoli and Carrasco-Urra write: 'We currently lack a mechanistic explanation for this unique phenomenon' but they hypothesise that 'plant volatiles trigger specific phenotypic changes in neighbouring vine leaves' (986). This is discussed in František Baluška and Stefano Mancuso, 'Vision in Plants via Plant-Specific Ocelli?', *Trends in Plant Science*, Vol. 21, No. 9 (September 2016): 727. The notion of plant 'ocelli' was first proposed by Gottlieb Haberlandt in 1905.

7 Baluška and Mancuso, 'Vision in Plants', 728.

8 Ernesto Gianoli, 'Eyes in the Chameleon Vine?', *Trends in Plants Science*, Vol. 22, No. 1 (1 January 2017): 4.

9 Fleur Daugey, *L'intelligence des plantes: Les découverts qui révolutionnent notre compréhension du monde végétal*, Les Éditions Ulmer, Paris, 2018, 38. Daugey is referring to Charles Darwin, assisted by Francis Darwin, *The Power of Movement in Plants*, John Murray, London, 1880.

10 Stefano Mancuso, *Brilliant Green: The Surprising History and Science of Plant Intelligence*, trans. Joan Benham, Island Press, Washington, 2015 [2013], 46.

11 Peter Wohlleben, *The Hidden Life of Trees: What They Feel, How They Communicate: Discoveries from a Secret World*, trans. Jane Billinghurst, William Collins, London, 2017 [2015], 148. He similarly claims that if plants can 'identify' (i.e. react in a certain way) to the saliva of particular insects (e.g. by emitting substances that attract the predators of these insects) 'they must also have a sense of taste' (9).

12 Daniel Chamowitz, *What a Plant Knows: A Field Guide to the Senses of Your Garden and Beyond*, Oneworld, London, 2012, 30, 16, 31, 27.

13 See Monica Gagliano, Stefano Mancuso and Daniel Robert, 'Towards Understanding Plant Bioacoustics', *Trends in Plant Science*, Vol. 17, No. 6 (June 2012): 323–5.

14 Monica Gagliano and Martial Depczynski, 'Tuned In: Plant Roots Use Sound to Locate Water', *Oecologica*, Vol. 184, No. 1 (2017): 151–60.

15 Daugey, *L'intelligence des plantes*, 42; Mancuso, *Brilliant Green*, 72. See also the BBC article http://www.bbc.com/earth/story/20170109-plants-can-see-hear-and-smell-and-respond (last accessed 14 January 2021) Chamovitz, despite a chapter entitled 'What a Plant Hears', concludes that plants are in fact 'deaf' (*What a Plant Knows*, 101–10).

16 Jacques Tassin, *À quoi pensent les plants?*, Odile Jacob, Paris, 2016, 59. See also 69.

17 See Frank W. Telewski, 'A Unified Hypothesis of Mechanoperception in Plants', *American Journal of Botany*, Vol. 93, No. 10 (2006): 1466–76.

18 For another example, see Marine Veits et al., 'Flowers Respond to Pollinator Sound within Minutes by Increasing Nectar Sugar Concentration', *Ecology Letters*, Vol. 22 (2019): 1483–92. Flowers exposed to the recorded sound of a flying bee or synthetic sounds at similar frequencies 'vibrated mechanically in response' and produced sweeter nectar within three minutes. This suggests, the authors write, that the flower 'serves as an auditory sensory organ… We hypothesise that the flower serves as an external "ear".' Once again, this conflates vibration with sound. For a critique of the equation of vibration and sound see Tassin, *À quoi pensent les plantes?*, 58–9; Florence Burgat, *Qu'est-ce qu'une plante? Essai sur la vie végétale*, Seuil, Paris, 2020, 90.

19 See Mauseth, *Botany*, 307–10.

20 See, for example, R. Zweifel and F. Zeugin, 'Ultrasonic Acoustic Emissions in Drought-stressed Trees – More than Signals from Cavitation?', *New Phytologist*, Vol. 179 (2008): 1070–9. Zweifel and Zeugin identify two possible origins for some of these ultrasonic emissions: 'the respiration and the metabolic growth activity of cambium or ray parenchyma cells' (1077).

21 Gagliano, Mancuso and Robert, 'Towards Understanding Plant Bioacoustics', 324–5.

22 Ibid., 325: 'multidisciplinary research is required for an effective exploration of the functional, ecological and ultimately evolutionary significance of acoustic communication in the life of plants.' Wohlleben (*The Hidden Life of Trees*, 15) glosses this as follows: as seedling roots grow towards a crackling at a certain frequency 'That means the grasses were registering this frequency, so it makes sense to say that they "heard" it.' See also Monica Gagliano, 'Green Symphonies: A Call for Studies on Acoustic Communication in Plants', *Behavioural Ecology*, Vol. 24, No. 4 (July–August 2013): 789–96.

23 One of the earliest studies to show this is Jan T. Baldwin and Jack C. Schultz, 'Rapid Changes in Tree Leaf Chemistry Induced by Damage: Evidence for Communication between Plants', *Science*, Vol. 221, No. 4607 (15 July 1983): 227–9. For a brief survey see Anthony Trewavas, *Plant Behaviour and Intelligence*, Oxford University Press, Oxford, 2014, 185–6.

24 Ian T. Baldwin, André Kessler, Rayko Halitschke, 'Volatile Signalling in Plant-plant Herbivore Interactions: What Is Real?', *Current Opinion in Plant Biology*, Vol. 5 (2002): 1–4, 1.

25 See, for example, Junji Takabayashi and Marcel Dicke, 'Plant-Carnivore Mutualism through Herbivore-Induced Carnivore Attractants', *Trends in Plant Science*, Vol. 1, No. 4 (1996): 109–13.

26 Baldwin, Kessler, Halitschke, 'Volatile Signalling in Plant-Plant Herbivore Interactions', 1.

27 In a later article Baldwin, Halitschke and Paschold ('Volatile Signalling in Plant-Plant Interactions: "Talking Trees" in the Genomics Era', *Science*,

Vol. 311 (2006): 812–15) incorporate the terms of the popular interpretation, albeit in quotations marks: Plants 'eavesdrop' on each other, some plants are 'deaf' or 'mute', and so on.

28 Mancuso, *Brilliant Green*, 54; Wohlleben, *The Hidden Life of Trees*, 249 (Wohlleben also has a chapter on 'The Language of Trees', 6–13); Quanta Magazine, 'The Secret Language of Plants: Striking Evidence that Plants Warn Each Other of Environmental Dangers Is Reviving a Once Ridiculed Field', https://www.quantamagazine.org/the-secret-language-of-plants-20131216/ [last accessed 16 January 2021]; BBC Earth, 'Plants talk to each other using an internet of fungus', http://www.bbc.co.uk/earth/story/20141111-plants-have-a-hidden-internet [last accessed 16 January 2021]; Guardian, 'Plants "talk to" each other through their roots: Scientists studying corn seedlings believe that they send signals under the soil, advising each other of the proximity of other plants', https://www.theguardian.com/science/2018/may/02/plants-talk-to-each-other-through-their-roots.

29 See Attenborough, *The Private Life of Plants*, 83.

30 Chamovitz, *What a Plant Knows*, 58. Jacques Tassin (*À Quoi pensent les plantes?*, 55) bluntly characterizes this kind of anthropomorphism (talking trees, especially) as 'travestying' the scientific literature 'to satisfy our fantasies of plants endowed with a mind [*esprit*] similar to ours'. But Macuso's work is part of the scientific literature.

31 Mancuso, *Brilliant Green*, 46. Richard Karban's, *Plant Sensing and Communication,* University of Chicago Press, Chicago and London, 2015, gives a good overview of its topic and is careful to avoid the kind of anthropomorphism we are discussing here (while being aware of the literature that does do that).

32 See, for example, Attenborough, *The Private Life of Plants*, 9; Michael Pollan, *The Botany of Desire: A Plant's Eye-View of the World*, Bloomsbury, London, 2001. Mancuso writes: 'When the absurd subjection of the plant world to the animal world finally comes to a halt, it will be possible to study plants – much more usefully – for their differences from animals, rather than their similarities to them' (*Brilliant Green*, 25–6). But we do not see that attempt in his *Brilliant Green*; quite the opposite, in fact.

33 Peter Tompkins and Christopher Bird, *The Secret Life of Plants: A Fascinating Account of the Physical, Emotional, and Spiritual Relations between Plants and Man*, Harper, New York, 1973. A documentary based on Tompkins and Bird's book, also called 'The Secret Life of Plants' (1979), features a soundtrack by Stevie Wonder (https://topdocumentaryfilms.com/the-secret-life-of-plants/). Michael Pollan discusses the efforts of contemporary plant scientists and plant advocates (himself included) to distance themselves from Tompkins and Bird in 'The Intelligent Plant', *The New Yorker*, 16 December 2013. Print edition (23 and 30 December 2013) https://www.newyorker.com/magazine/2013/12/23/the-intelligent-plant

34   Chamovitz, *What a Plant Knows*, 6.

35   Mancuso, *Brilliant Green*, 46. Confusingly, in a later book Mancuso also says that 'Plants have nothing in common with us'. Stefano Mancuso, *The Revolutionary Genius of Plants: A New Understanding of Plant Behaviour and Intelligence*, trans. Vanessa di Stefano, Atria Books, NY etc, 2018 [2017], 71.

36   Daugey, *L'intelligence des plantes*, 53.

37   It is curious that in a popular scientific book the author says the kind of thing that would not pass muster in a high school biology class. How do trees feel about fungal threads growing into their roots? Wohlleben (*The Hidden Life of Trees*, 51) writes that there is no research into whether this is painful or not, 'but as it is something the tree wants, I imagine it gives rise to positive feelings'.

38   Ruth Kassinger, *A Garden of Marvels: How We Discovered that Flowers Have Sex, Leaves Eat Air, and Other Secrets of Plants*, William Morrow, New York, 2012, 304.

39   Michael Allaby, *Plant Love: The Scandalous Truth about the Sex Life of Plants*, Filbert Press, 2016, 63. Allaby's book reads like an updated print version of the infamous television documentary 'Sexual Encounters of the Floral Kind: An Investigation into the Extraordinary Sex Life of Plants', from the series The World About Us, Oxford Scientific Films, written and directed by Bill Travers, narrated by Freddie Jones, 1980 (https://www.youtube.com/watch?v=4b35nYXVNKI).

40   In a chapter entitled 'Sex, Sex, Sex', Apt Russell writes (*Anatomy of a Rose: The Secret Life of Flowers*, William Heinemann, London, 2001, 49): 'The Jack-in-the-pulpit is considering a sex change. The violets have a secret. The dandelion is smug. The daffodils are obsessive. The orchid is *finally* satisfied, having produced over a million seeds. The bellflower is *not* satisfied and is slowly bending its stigma in order to reach its own pollen. The pansies wait expectantly, their vulviform faces lifted to the sky. The evening primrose is interested in one thing and one thing only. A stroll through the garden is almost embarrassing'.

41   See, for example, Mancuso, *Brilliant Green*, 126–7; Daugey, *L'intelligence des plantes*, 13. The full scope of Trewavas's position and its ultimately philosophical claims only really become clear with the extended spread of discussion in his *Plant Behaviour and Intelligence* (2014), which postdates some of the popular literature (including Chamovitz's *What a Plant Knows* and Mancuso's *Brilliant Green*) and is not discussed in any detail in any of it.

42   Trewavas, 'Aspects of Plant Intelligence', *Annals of Botany*, Vol. 92, No. 1–20 (2003): 1. See also Trewavas, 'The Foundations of Plant Intelligence', *Interface Focus*, Vol. 7 (2017): 2.

43   See, for example, Trewavas, *Plant Behaviour and Intelligence*, 1, 74.

44 See Trewavas, *Plant Behaviour and Intelligence*, 245–6. Trewavas refers to Loeb's books *The Mechanistic Conception of Life* (1912) and *Forced Movements, Tropisms and Animal Conduct* (1918).

45 The point here is that, whereas arguments exist to the effect that we need to think human intelligence less reductively, the reductive version of intelligence seems to dominate in objections to the idea of plant intelligence. The idea that the behaviour of plants (or indeed any living being) could be reduced to reflex reaction is basically implausible, according to Trewavas. Granting that an organism in an identical molecular and physiological state, with an identical stimulus in an identical environment, might produce an identical response, the fact is that no such identical circumstances can plausibly to be postulated. Further, any organism limited to a set of 'pre-programmed' responses (even presuming that to be possible) would not be able to adapt and would not survive. Paco Calvo, Monica Gagliano, Gustavo M. Souza and Anthony Trewavas, 'Plants Are Intelligent, Here's How', *Annals of Botany*, Vol. 125 (2020): 11–28, offer a detailed and compelling argument for understanding plant intelligence in relation to survival and adaptation.

46 Trewavas, *Plant Behaviour and Intelligence*, 196.

47 Chamovitz, *What a Plant Knows*, 168.

48 Trewavas, *Plant Behaviour and Intelligence*, 193. Marder (*Grafts: Writings on Plants*, Univocal, Minneapolis, 2016, 72–3) endorses this view of plant intelligence.

49 Trewavas, *Plant Behaviour and Intelligence*, 23.

50 Ibid., 233.

51 Ibid., 23.

52 From this point of view should we then have to say that viruses are alive – perhaps not at the level of each particular virion but at the level of the species? There may, of course, be other forms of (non-biological) intelligence, notably artificial intelligence or machine intelligence, but the claims here concern specifically biological intelligence. Swarm intelligence is a form of biological intelligence. The effective identification of intelligence and life also means that the idea of 'minimal intelligence', as discussed in Paco Calvo and František Baluška ('Conditions for Minimal Intelligence Across Eukaryota: A Cognitive Science Perspective', *Frontiers in Psychology*, 3 September 2015), is odd. 'Minimal' (as opposed to 'full blown') intelligence would be minimal (as opposed to full blown) life.

53 Although he thinks we cannot identify a 'mark of intelligence', the way that Calvo presents his questions about plant intelligence sometimes suggests that it is something that an organism may or may not have, and that it may be inferred from a constellation of behaviours. See Paco Calvo, 'The Philosophy of Plant Neurobiology', *Synthese*, Vol. 193 (2016): 1323–43, 1325.

54 Quentin Hiernaux ('Pourquoi et comment philosopher sur le vegetal?', 17) reminds us, indeed, that the concept of 'intelligence' is primarily philosophical, not scientific.

55 Chamovitz's objection to Trewavas's idea of plant intelligence is in fact an objection to something else – the idea of 'plant neurobiology', a new field of research that arose in the early twenty-first century. Based on claimed analogies between aspects of neural functioning in animals and aspects of plant physiology (concerning electrical signalling, homologous molecules, especially glutamate, and 'cell-cell transport' processes) a group of scientists (including, notably, Stefano Mancuso and František Baluška) coined the term 'plant neurobiology' to describe the investigation of 'the structure of the information network that exists within plants' (Eric D. Brenner, Rainer Stahlberg, Stefano Mancuso, Jorge Vivanco, František Baluška and Elizabeth Van Volkenburgh, 'Plant Neurobiology: An Integrated View of Plant Signalling', *Trends in Plant Science*, Vol. 11, No. 8 (2006): 413–19, 413). Within a matter of months, in a letter signed by thirty-three scientists, the idea of plant neurobiology was criticized for its lack of evidential basis and because – so these scientists argued – there was nothing to be gained for plant science from thinking in neurobiological terms (Amadeo Alpi et al., 'Plant Neurobiology: No Brain, No Gain?', *Trends in Plant Science*, Vol. 12, No. 4 (2007): 134–5). Except for attempts to locate the neural analogues in specific parts of the plant, the research conducted under the banner of plant neurobiology is basically research into plant behaviour and intelligence. Trewavas signals as much, when he defends Brenner et al. by arguing that 'plant neurobiology' is just a metaphor. See Trewavas, 'Reply to Alpi et al: Plant Neurobiology – Metaphors Have Value', *Trends in Plant Science*, Vol. 12, No. 6 (June 2007): 231–3; Trewavas, *Plant Behaviour and Intelligence*, 150, 155. Jacques Tassin's objection to the idea of plant intelligence (*À quoi pensent les plantes?*, 40) is similarly, effectively, an objection to plant neurobiology, or to the idea that the claim for plant intelligence can be based on neuronal analogies. Tassin also denies (41) that plant intelligence can be deduced from plant movement. But as we have shown, the idea of plant intelligence in Trewavas's later work is quite different to either of these claims. Hallé's objection to the idea of plant intelligence (*Éloge de la plante*, 320–1) also misses its target. Plants are not intelligent, he writes; because their responses are programmed, they have no choice. But intelligence does not necessarily imply choice (as discussed below) and Trewavas argues convincingly that no organism could be 'programmed' to respond in a particular way to every possible, unique situation.

56 Trewavas, 'What Is Plant Behaviour?', 606, 610 and passim.

57 See the 2009 BBC series 'Life', Episode 9, 'Plants' and the 2022 BBC series 'Green Planet', Episode 2.

58 For example, Attenborough, *The Private Life of Plants*, 7; 'Green Planet', Episode 2; Chamovitz, *What a Plant Knows*, 71; Daugey, *L'intelligence des* *plantes*, 108, 112, 115. Tassin (*À quoi pensent les plantes*, 80–2) objects to the use of the term 'memory' in relation to plants because for him memory requires an individual who can compare past and present in the reproduction of a past representation. However, in counting it as a metaphor, it may be what he really objects to is the anthropomorphic interpretation of plant 'memory', not the (insufficiently careful) use of the term itself.

59 See, for example, Minoru Ueda and Yoko Nakamura, 'Metabolites Involved in Plant Movement and "Memory": Nyctinasty of Legumes and Trap Movement in the Venus Flytrap', *Natural Product Reports*, Vol. 23 (2006): 548–57, especially 556–7.

60 See, for example, Trewavas, 'What Is Plant Behaviour?', 611–2.

61 Wohlleben claims (*The Hidden Life of Trees*, 149) that temperature sensitivity in trees 'proves… that trees must have a memory. How else could they inwardly compare day lengths or count warm days?' Authors like Chamovitz (*What a Plant Knows*, 158, 161) speak freely of the plant 'remembering' events, when we do not speak of machine learning in this way.

62 Trewavas, *Plant Behaviour and Intelligence*, 75.

63 Ibid., 27.

64 Ibid., 90.

65 See, for example, Georg Toepfer, 'Teleology and Its Constitutive Role for Biology as the Science of Organized Systems in Nature', *Studies in History and Philosophy of Biological and Biomedical Sciences*, Vol. 43 (2012): 113–19; and other articles in this issue.

66 Trewavas, *Plant Behaviour and Intelligence*, 90. We see the same conflation in Calvo et al., 'Plants Are Intelligent, Here's How', 19 and in Ken Thompson, *Darwin's Most Wonderful Plants: Darwin's Botany Today*, Profile Books, London, 2018, 86: 'Plants display an intentional, purposeful ability to solve problems, and that's as good a definition of intelligent behaviour as any.'

67 See Immanuel Kant, *Critique of Pure Reason*, trans. Paul Guyer and Allan W. Wood, Cambridge University Press, Cambridge, 1997 [1781/1787], 411–17 (A341–351). In the first 'paralogism of pure reason' Kant argues that, in the history of philosophy and theology, the legitimate claim (first premise) that the 'I' is the absolute grammatical subject of judgement (i.e. that it cannot be a predicate of a judgement) is conflated with the claim (effectively, the second premise) that an absolute subject is a substance (another, different meaning of the concept of a subject), to give the unwarranted conclusion that the I, as a thinking being, is substance. The problem is that the same word ('subject') is used in different senses in the two claims but treated as if it meant the same in both.

68 Trewavas, *Plant Behaviour and Intelligence*, 51, 77. Trewavas also speaks of animals having made evolutionary choices; see, for example, 185.

69 Mancuso, *Brilliant Green*, 32, 33, 35, 138, 109. The strongest (not to say most scathing) criticism of Mancuso comes from Florence Burgat (*Que est-ce qu'une plante?*, 71). Commenting on the attribution of decision-making to plants she asks, incredulously, 'To what ends do these authors sacrifice all scientific rigour? Do they have faith in the comparisons that they dare to make? Do they believe in their theories?' The language of decision is echoed in the description of the recent (2021) online exhibition from the Camden Arts Centre in London, 'The Botanical Mind: Art, Mysticism and the Cosmic Tree' (https://camdenartcentre.org/the-botanical-mind-online/): 'Millions of years ago plants chose to forego mobility in favour of a life rooted in place, embedded in a particular context or environment.'

70 Marder, *Plant Thinking*, 9, writes of the attempt to describe the world 'from the standpoint of the plant itself'. This is, of course, impossible for a non-plant, but as Tassin points out (*À quoi pensent les plants?*, 11), it is a 'salutary position' from which to begin the investigation into the ontology of plant being.

71 Heidegger famously describes the task of his *Being and Time* (1927) as 'The task of destroying [*die Aufgabe einer Destruktion*] the history of ontology'. *Being and Time*, trans. John Macquarrie and Edward Robinson, Basil Blackwell, Oxford, 1962, H19, 41. Derrida's '*déconstruction*' is a philosophical translation of Heidegger's '*Destruktion*'.

72 Marder, *Plant Thinking*, 11, 90; see also 53, 83 and Marder in Luce Irigaray and Michael Marder, *Through Vegetal Being: Two Philosophical Perspectives*, Columbia University Press, NY, 2016, 112.

73 Tassin, *À quoi pensent les plants?*, 7, 102,

74 Ibid., 121. See also Marder, *Grafts*, 74, where he envisages a 'creative symbiosis of philosophy and botany'.

75 See Tobias Cheung, 'From the Organism of a Body to the Body of an Organism: Occurrence and Meaning of the Word "Organism" from the Seventeenth to the Nineteenth Centuries', *British Journal for the History of Science*, Vol. 39, No. 3 (September 2006): 319–39.

76 See, for example, Mancuso, *Brilliant Green*, 34, 126; Mancuso, *The Revolutionary Genius of Plants*, 72–3; Marder, *Plant Thinking*, 84; Jacques Tassin, *Penser comme un arbre*, Odile Jacob, Paris, 2018, 40; Elaine P. Miller, *The Vegetative Soul: From Philosophy of Nature to Subjectivity in the Feminine*, SUNY Press, New York, 2002, 8.

77 Cf. Tassin, *Penser comme un arbre*, 54.

78 Embryonic stem cells are pluripotent, but not totipotent. Once the zygote has divided a few times to become the blastocyte, the inner cell mass (from which the embryonic stem cells derive) produces all parts of the embryo but not the placenta. See Nessa Carey, *The Epigenetics Revolution: How Modern Biology Is Rewriting Our Understanding of Genetics, Disease and Inheritance*, Icon, London, 2012, 27–8.

79 There are some exceptions, including some healing processes and cancer.

80 See Hallé, *Éloge de la plante*, 144–8.

81 Hallé, *Éloge de la plante*, 100.

82 Tassin, *À quoi pensent les plantes?*, 20, 22.

83 Hallé, *Éloge de la plante*, 42–51. Maximization of surface area is maximization of photosynthetic opportunity. Not all plants, of course, are photosynthetic – that is, autotrophic. Parasitic plants can be complete heterotrophs, but they still exhibit morphological plasticity.

84 Hallé, *Éloge de la plante*, 197–8.

85 See Francis Hallé, *Tropical Trees and Forests: An Architectural Analysis*, Springer, New York, 1978.

86 As a form of woodland management this is called 'coppicing'; see Oliver Rackham, *Woodlands*, William Collins, London, 2015 [2006], 16–17.

87 Tassin, *Penser comme un arbre*, 60, 22.

88 See Hallé, *La vie des arbres*, Montrouge, Bayard, 22.

89 See Tassin, *À quoi pensent les plantes*, 25; Hallé, *Éloge de la plante*, 126.

90 The woodland ecologist and historian Oliver Rackham refers to these and other aspects of plant life when he asks, in relation to trees, what is normal? 'The public has been trained', he says, 'by 250 years of Enlightenment to expect trees to be upright and single-stemmed, not to have dead branches, not to be rotten, to come into leaf and lose their leaves at the right time of year, and to die of old age'. But this does not 'agree with the agenda of trees', as he puts it. Rackham, *Woodlands*, 342.

91 Hallé, *Éloge de la plante*, 279. Heidegger (*Being and Time*, H246–7, 290–1) reserves the existential possibility of 'dying' for Dasein, that entity which we ourselves are, entities whose being is an issue for them and who have the possibility of an authentic anticipation of death. According to Heidegger, animals are subject to death but they do not 'die'; like plants, they 'perish'.

92 Hallé, *Éloge de la plante*, 116–18.

93 Hallé, *Éloge de la plante*, 120.

94 John L. Harper, *Population Biology of Plants*, Academic Press, London etc, 1977, 20.

95 Merlin Sheldrake (*Entangled Life: How Fungi Make Our Worlds, Change Our Minds, and Shape Our Futures*, The Bodley Head, London, 2020, 24) also notes how fungal life and its relations 'roll us towards the edge of many questions' (24), including that of the compass of the notion of the individual (19).

96 See, for example, Barbara McClintock, 'The Origin and Behaviour of Mutable Loci in Maize', *Proceedings of the National Academy of Sciences of the United States of America*, Vol. 36, No. 6: 344–55.

97 Hallé, *Éloge de la plante*, 211.

98 See Hallé, *Éloge de la plante*, 238. The same is true of some sponges; this is one more reason why Hallé argues that a plant is more like a colony than an individual. Hallé also (203–13) suggests that somatic mutation may be understood as an evolutionary strategy.

99 See Hallé, *Éloge de la plante*, 118–19.

100 See Hallé, *Éloge de la plante*, 263–4. Ironically the genealogical tree model represents animal phylogeny, but cannot deal with aspects of plant evolution, for example, hybridity.

101 Because of its modular structure, the plant, for Hallé (e.g. *Éloge de la plante*, 111) is better thought as a colony, rather than an individual. He asks, does 'unity' mean the same thing when we speak of plants and when we speak of animals? Lenn Jerling ('Are Plants and Animals Alike? A Note on Evolutionary Plant Population Ecology', *Oikos*, Vol. 45, No. 1 (August 1985): 150–3) writes that 'the most appropriate dividing line is not between plants and animals but between modular and unitary organisms'. In this short article Jerling discusses many of the problems raised by plant philosophy, including that of the individual.

102 On 'self-awareness' in plants see Trewavas, *Plant Behaviour and Intelligence*, 181–3. The philosopher Matthew Hall – who affirms, as a starting point for a different kind of plant philosophy, that plants should be recognized as 'subjects deserving of respect as other-than-human persons' – argues for a conception of the 'plant self', because allogenetic mechanisms show that 'plants also recognise themselves as integrated beings'. But are they 'integrated beings'? Hall relies on Trewavas's work on 'self-recognition' and does not consider the problem that plant life poses to the category of the individual. He too criticizes the zoocentrism of the approach to plants in the Western philosophical tradition, but again this means the denegation of plants and the characterization of them as passive, inert etc., not the problem of the animal model. In a chapter entitled 'Bridging the Gulf' he conflates 'the constructed inferiority' of plants with the 'rendering of plants as radically different', arguing that 'this positioning of plants as radically different... is contradicted by an overwhelming body of evidence that has been accruing in the botanical sciences' (137). For Hall then, we revalue plants by showing that they are not so different from us after all. But in so doing Hall does not avoid the zoocentrism of the animal model, as his use of the category of 'person' arguably shows. See Hall, *Plants as Persons: A Philosophical Botany*, SUNY Press, New York, 2011, 13, 149. Referring to Hall, Marder (*Plant Thinking*, 55) writes, 'It is neither necessary nor helpful to insist... on the need to attribute to vegetal beings those features, like autonomy or even personhood, philosophers have traditionally considered as respect-worthy'. However, Hall also draws on non-Western sources and considers the possibilities raised by 'indigenous animism', which complicates matters considerably (99ff). We will return to Hall when we discuss Eduardo Viveiros de Castro's account of Amerindian perspectivism in Chapter 6.

103 In *Grafts* (38) Marder writes that plant thinking is not a 'species-thought' and 'plants' are not just the category otherwise known as the biological kingdom Plantae: '"plants" stands for a tendency of living and thinking that promotes growth, decay and metamorphosis.' In their *Radical Botany: Plants and Speculative Fiction,* Fordham University Press, New York, 2020, Natania Meeker and Antónia Szabari similarly see plants as 'the model for all animate life' (1).

104 Marder, *Plant Thinking*, 5. On vegetal ontology see ibid., 9.

105 Ibid., 18. We discuss Aristotle's notion of the 'nutritive' or 'vegetative' soul at greater length in Chapter 2.

106 Marder, *Plant Thinking*, 19.

107 Ibid., 51.

108 The notion of the 'transcendental synthesis of the "I think"' is part of Immanuel Kant's outstandingly influential philosophy – for some the moment of a new phase in Western philosophy. Here 'transcendental' means that which is a condition of possibility for something else; the 'I think' that must be able to accompany all of my representations is the condition of possibility for a unified 'self'. See Kant, *Critique of Pure Reason*, 246–8 (B132–4): 'The I think must be able to accompany all my representations… otherwise I would have as multicoloured, diverse a self as I have representations of which I am conscious.'

109 Marder, *Plant Thinking*, 164, 165, 170. Kant's discussion of the 'I think' is in fact less 'self-identical' than Marder's opposition might here suggest – but perhaps, ultimately that is Marder's point.

110 Marder, *Plant Thinking*, 167, 173. In his contribution to *through Vegetal Being* (124) Marder writes that 'The core argument of *Plant Thinking*' was 'that humans share intelligence and life with plants, from which much of our own thinking and being derive'.

111 Marder, *Plant Thinking*, 43. Meeker and Szabari (*Radical Botany*, 1) also see the 'way of being' of the plant as 'profoundly generative for human thought and practice.' For them plants 'inspire speculative activity in our efforts to think with them' (17) because plant life 'resists incorporation into human categories; the study of plants becomes, in this context, an engine of social critique and speculation about possible futures' (16). *Radical Botany* is a series of investigations into the tradition of speculative fiction where the plant figures as a way of exploring the limitations of human perception (23) and alternative ways of understanding the human relation to the environment.

112 Emanuele Coccia, *The Life of Plants: A Metaphysics of Mixture*, trans. Dylan J. Montanari, Polity, Cambridge, 2019 [2017], 5, 10.

113 Ibid., 5, 66, 32.

114 Coccia claims, perhaps a little grandiosely, that, on the basis of his 'metaphysics of mixture', 'a new geometry must be thought out', 'Physics – the science of nature should then be completely rewritten.' Ibid., 69, 71.

115 Tassin, *Penser comme un arbre*, 58, 127. 'To accept the tree in its quintessence is to surrender to the vertigo that is inherent in the recognition of an *other* form of life' (126–7).

116 This is the case, for example, with some insects, like bees. Some unfertilized haploid eggs develop into sterile 'worker' bees'.

117 The animal life cycle involves one 'generation' or 'phase' which is generally understood to have four stages: birth, growth, reproduction and death. Plants are not born; their growth may be vegetative reproduction itself. The two generations (sporophyte and gametophyte) are part of sexual reproduction, where it occurs; they die, but not in the way of animals.

Chapter 2

1 Lynn Margulis and Dorion Sagan, *Origins of Sex: Three Billion Years of Genetic Recombination*, Yale University Press, New Haven, 1986, 195.

2 A. G. Morton, *History of Botanical Science: An Account of the Development of Botany from Ancient Times to the Present Day*, Academic Press, London etc, 1981, 184, 238. See also Fara, *Sex, Botany and Empire*, 21: 'Surprising though it might seem, it had been nearly the end of the 17th century before naturalists realised that plants reproduce sexually.'

3 Morton, *History of Botanical Science*, 242.

4 Scott Atran (*Cognitive Foundations of Natural History: Towards an Anthropology of Science*, Cambridge University Press, Cambridge, 1990, 85) notes that when historical figures are treated as 'precursors' with regard to the treatment of problems that agitate contemporary figures 'the historical past often appears as an imperfect prelude to the modern present'. And so it is with narratives of the development of the theory of the sexuality of plants.

5 Taiz and Taiz, *Flora Unveiled*, vii. According to Taiz and Taiz (ibid.) 'our approach to the history of the discovery of sex in plants is both "whiggish" and contextual'.

6 Taiz and Taiz, *Flora Unveiled*, 214.

7 These analogies are not absent in, for example, the tradition of medieval herbals, and they form a major part of the botanicals concerned with the doctrine of signatures. Our discussion here, however, is restricted to those texts in the tradition of Aristotelian botany that are important for the history of the ascription of sex to plants.

8 Aristotle, *Parts of Animals*, trans. D. M. Balme, Clarendon Press, Oxford, 1992, 644a17–23, 15–16. On Aristotle's use of the logical categories of genus (*genos*) and species (*eidos*) and their relation to the modern taxonomical categories, see Stella Sandford, 'From Aristotle to Contemporary Biological Classification: What Kind of Category Is "Sex"?', *Redescriptions: Political Thought, Conceptual History and Feminist Theory*, Vol. 22, No. 1 (2019), 4–17.

9 The parts of the animal body are 'for the sake of the function in relation to which each [part] has naturally grown'. Aristotle, *Parts of Animals*, 645b20–21, 19.

10 Aristotle, *Parts of Animals*, 645b7–11, 19.

11 See Mary B. Hesse, *Models and Analogies in Science*, University of Notre Dame Press, Indiana, 1966, 134, 140. We will return to analogy and its distinction from homology in modern biology in the Epilogue to this book.

12 Aristotle, *History of Animals*, trans. D'Arcy Wentworth Thompson, in *The Complete Works of Aristotle*, Vol. I, Princeton University Press, Princeton, NJ, 1984, II, 1, 18–20, 792.

13 In *Parts of Animals* (trans. W. Ogle in *The Complete Works of Aristotle*, Vol. I, IV, 5, 13–16, 1063), having described the mouth parts of various sea creatures, Aristotle makes the point explicitly: 'Such then, is the structure of the parts that minister to nutrition and which every animal must necessarily possess. But besides these organs it is quite plain that in every animal there must be some part or other which shall be analogous to what in sanguineous animals is the presiding seat of sensation.' In animals this 'principle of sensation' is located in the heart; see Aristotle, 'On Youth, Old Age, Life and Death, and Respiration', trans. G. R. T. Ross, in *The Complete Works of Aristotle*, Vol. I, 3–4, 747–8. Aristotle also claims that what some people calls the '*mytis*' is the corresponding part in the Cephalopoda and the Crustacea (*Parts of Animals*, IV, 5, 16–20).

Atran (*Cognitive Foundations of Natural History*, 105) writes that Aristotle's 'comparative morphology' allowed him to compare animals and man systematically, and that 'it is this systematic effort to exhaustively compare the structures and functions of other animals to man that led Aristotle to introduce the concept of *analogy* to unify all life. Henceforth it would be possible not only to scrutinize the hitherto inscrutable invertebrates; even plants might be ultimately assimilable to the same systematic framework'.

14 Hesse, *Models and Analogies in Science*, 134.

15 Ibid. Aristotle uses the same example in *Posterior Analytics* (trans. Jonathan Barnes, in *The Complete Works of Aristotle*, Vol. I) where he writes that we can excerpt features from various animals, for example, 'in virtue of analogy; for you cannot get one identical thing which pounce and spine and bone should be called; but there will be things that follow them too, as though there were some single nature of this sort' (II, 14, 98a20–22, 162), As Hesse (*Models and Analogies in Science*, 133) says, it is thus sometimes in virtue of analogy that a common name can be coined.

16 Aristotle, *De Anima*, trans. Christopher Shields, Clarendon Press, Oxford, 2016, II, 1, 412a25–7, 22. As Shields (171) explains in his commentary: one may have actualized one's potential to attain knowledge of something but not actually be using that knowledge while sleeping; when awake that first actualization can become the second actualization of contemplating.

17 Aristotle, *Metaphysics*, trans. W. D. Ross, in *The Complete Works of Aristotle*, Vol. II, IX, 6, 1048 b1–9, 1655; IX, 5, 1047b35–6, 1654.

18 'As for the underlying nature, it must be grasped by analogy. As bronze stands to a statue, or wood to a bed, or [the matter and] the formless before it acquires a form to anything else which has a definite form, so this stands to a reality, to a this thing here, to what is' (*Physics*, trans. W. Charlton, Clarendon Press, Oxford, 1970, I 7, 191a8–12, 18).

19 Aristotle, *Metaphysics*, V, 6, 1016b31–1017a3, 1605.

20 Hesse, *Models and Analogies in Science*, 141, 149. See also G. E. R. Lloyd, *Analogical Investigations: Historical and Cross-cultural Perspectives on Human Reasoning*, Cambridge University Press, Cambridge, 2015, 79–80.

21 Most commentators agree that in Aristotle's zoological works comparisons between genera are analogous, whereas those within genera are comparisons of the more or less. See, for example, A. Gotthelf's, Introduction to *History of Animals*, Loeb, Harvard University Press, Cambridge MA and London, 1991, 16.

22 *History of Animals*, 588b4–13, 922. The genus of testaceans (roughly, seashells) is particularly ambivalent according to Aristotle, resembling plants to the extent that they do not (so he seems to think) move (*History of Animals*, 588b16–18, 922). See also Aristotle, *Generation of Animals* (trans. A. Platt, in *The Complete Works of Aristotle*, Vol. II, III, 11, 761a15–16, 1178: 'compared with animals they [testacea] resemble plants, compared with plants they resemble animals'.

23 Aristotle, *De Anima*, II, 1, 412a20–1, 412a14, 22.

24 Ibid., II, 4, 415b15–22, 29. Earlier he had written: 'Hence, if it is necessary to say something which is common to every soul, it would be that the soul is the first actuality of an organic natural body' (II, 1, 412b5–6, 23).

25 Ibid., 413a22–25, 24, 413a26–7. This discussion shows that it was not taken for granted that plants are living things.

26 Aristotle, 'Sense and Sensibilia', in J. I. Beare, trans., *The Complete Works of Aristotle*, Vol. I, 436b1 11–13, 693: 'Sensation must, indeed, be attributed to all animals as such, for by its presence or absence we distinguish between what is and what is not an animal.' Aristotle makes the same point in 'On Youth, Old Age, Life and Death, and Respiration', 467b22–25,

745: 'a thing need not, though alive, be animal; for plants live without having sensation, and it is by sensation that we distinguish animal from what is not animal'.

27 Aristotle, *De Anima*, II, 3, 414b17–18, 27.

28 Ibid., II, 4, 415a25–7, 29. The classical Greek word for 'plant', *to phuton*, means quite generally 'that which has grown' and is connected to the verb *phuō*, to bring forth, produce, put forth (leaves and so on) and to grow. When used of men (in the specific sense of males, it seems) it means to beget, engender, generate. The vegetable principle is then the principle of growth and generation.

29 Ibid., II, 1, 412b3–4, 23.

30 See, for example, *Parts of Animals*, IV 686b10 27–687a2, 1071; 'On Youth, Old Age, Life and Death, and Respiration', 468a1–11, 745–6: 'That part where food enters we call upper… while downward is that part by which the primary excrement is discharged… Plants are the reverse of animals in this respect… owing to their being stationary and drawing their sustenance from the ground, the upper part must always be down; for there is a correspondence between the roots of a plant and what is called the mouth in animals.' In *Generation of Animals,* II4, 740b9–10, 1149, Aristotle also compares the umbilicus to the root. In *Parts of Animals*, II, 3 650a5–6, 650a23–5, 1012, Aristotle says that in animals the mouth takes in 'unconcocted' food, which is then concocted in the stomach with the aid of natural heat, but the food that plants get from the earth by means of roots is already concocted, 'which is the reason why plants produce no excrement, the earth and its heat serving them in the place of a stomach'. The next line reverses the analogy, calling the stomachal sac of the animal 'as it were an internal substitute for earth'. Further, if the stomach and intestines represent the ground there must be in animals channels like roots through which the nutriment passes (v).

31 Aristotle, *Parts of Animals*, II, 10, 656a3, 1022.

32 Aristotle, *History of Animals*, V, 1, 539a16–25, 853.

33 Aristotle, *Generation of Animals*, I, 4, 717a21–3, 1114; II, 1, 733b24–5, 1138. Here 'semen' is a generic word for seed, and the female menses is sometimes said to be 'semen' in this generic sense. The more specific word for male semen is *gonē*.

34 See, for example, Taiz and Taiz, *Flora Unveiled*, 214: 'despite their considerable resources, neither Aristotle nor Theophrastus succeeded in advancing the problem of sex in plants much beyond what ancient Mesopotamians had learned through their mastery of artificial pollination in dioecious date palms and the use of the caprifig and fig wasps to pollinate edible figs. The failure of these two luminaries to recognize the more general role of pollination in hermaphroditic flowers has long puzzled historians of botany'. As we argue elsewhere, one of the things that makes

such a view anachronistic is the fact that there is no distinct concept (or word) for 'sex' in classical Greek. See Sandford, 'From Aristotle to Contemporary Biological Classification: What Kind of Category Is "Sex"?'

35 Aristotle, *Generation of Animals*, I, 2, 716a 14–31, 1113.

36 Ibid., II, 5, 741a16–17, 1150. Later the female role is wholly re-described in terms of her *incapacity* in relation to this male power: 'the male and the female are distinguished by a certain capacity and incapacity. (For the male is that which can concoct and form and discharge a semen carrying with it the principle of form... the first moving cause)'. See also IV, 11, 765a31–38, 1184. Many commentators note, as Jessica Gelber puts it ('Females in Aristotle's Embryology', in Andrea Falcon and David Lefebvre, eds., *Aristotle's Generation of Animals: A Critical Guide*, Cambridge University Press, Cambridge, 2018, 173), that 'male and female are primarily differentiated by the ability each has to concoct residues and make their respective reproductive contributions'. See, for example, Johannes Morsink, *Aristotle on the Generation of Animals: A Philosophical Study*, University Press of America, Lanham NY and London, 1982, 135; Karen M. Nielsen, 'The Private Parts of Animals: Aristotle on the Teleology of Sexual Difference', *Phronesis*, Vol. 53 (2008): 373–405, 382.

37 Aristotle, *Generation of Animals*, II, 5 741a 8–16, 1150, translation amended.

38 François Delaporte (*Nature's Second Kingdom: Explorations of Vegetality in the Eighteenth Century*, trans. Arthur Goldhammer, MIT Press, Cambridge MA, 1982 [1979], 19–20) notes the disparity between Aristotle's intuition that plants, like animals, are living things and 'the very rudimentary structure that he attributed to plants. This disparity was apparently the result of a compromise. For it is clearly impossible to reconcile the contradictory requirements of trying to explain the plant kingdom in terms of the animal while respecting the hierarchy of functional principles. Nutritive and generative principles are appropriate to both plants and animals it is true, but the sensitive and locomotive principles guide the actions of animals alone. Nutrition in the animal therefore involves the existence of a sense of hunger, a faculty of choice. Generation, for its part, requires a locomotive function in distinct individuals; sexual union presupposes separation. Thus it is scarcely possible to conceive of organs and functions in plants dependent on the effects of the sensitive and locomotive principles'.

39 Aristotle, *History of Animals*, I, 3 489a12, 778; *Generation of Animals*, I, 1 715a20–1, 1111.

40 Aristotle, *Generation of Animals*, III, 11 762b1–5, 1178.

41 Ibid., I, 2, 715b18–19, 1112. Later he elaborates that 'compared with animals... [testacea] resemble plants, compared with plants they resemble animals... plants and testacea are analogous... since the object of testacea is to be in such a relation to water as plants are to earth, as if plants were, so to say, land-shell fish, shell-fish water plants' (III, 11, 761a15–16, 1178). Elsewhere he says that the plant-like character of the urchin, for example, is evident in its being upside down. Noting that it must have some part that serves the nutritive capacity, Aristotle says: 'The urchin has what we may call its head and its mouth down below, and a place for the issue of the residuum up above' (*History of Animals*, IV 5 19–20, 841).

42 *Generation of Animals*, III, 11, 762a 24–35, 1180. In modern botany reproduction by 'budding off alongside' is one of the methods of what is still known as 'vegetative' reproduction.

43 Aristotle, *De Anima*, II, 1, 413a22–25, 24; *Generation of Animals*, III, 11 762b 3–4, 1180.

44 Aristotle, *Generation of Animals*, IV, 1, 763b21–5, 1182, translation amended.

45 Ibid., III, 10 759b28–30, 1176; III, 11, 762b6–12, 1180.

46 Ibid., I, 23, 730b33–731a4, 1134. The point is reiterated at II, I, 732a1–3, 1136, where the principle of men, animals and plants is 'the male and the female'.

47 Ibid., IV, 1, 764b36–765a1, 1183.

48 Ibid., I, 2, 715b20–25, 1112.

49 This is the case, for example, when Robert Mayhew says that, for Aristotle, plants do not need sexes because they only have the nutritive soul (*The Female in Aristotle's Biology: Reason or Rationalization*, University of Chicago Press, Chicago and London, 2004, 45).

50 Theophrastus, *Enquiry into Plants*, trans. Arthur Hort, Loeb, Harvard University Press, Cambridge MA and London, 1916, I, 1, 3.

51 Ibid., I, 1, 5.

52 Ibid., I, 2, 19. See also I, 1, 7 and I, 1, 9. Agnes Arber notes the extent to which early analogical reasoning and the belief in the equivalence of animal and plant bodies has left its mark on contemporary botanical vocabulary (Arber, 'Analogy in the History of Science', in M. F. Ashley Montagu, ed., *Studies and Essays in the History of Science and Learning, Offered in Homage to George Sarton*, Henry Schuman, NY, 1944, 228).

53 Theophrastus, *Enquiry into Plants*, I, 1, 3. Though, of course, the relation between the inter-uterine embryo/foetus and the maternal parent is a complex philosophical problem.

54 Ibid., I, 13, 95. The example here is the date palm.

55 Ibid., III, 3, 177. It is also said that in trees in which both male and female bear fruit the female has more and fairer fruit – however, some invert the names and call these trees male. Theophrastus also says that what some distinguish as different kinds (what we would call 'species') others call male and female versions of the same thing (III, 9, 213). For example, what some call 'cork oak' others call 'female kermes oak' (III, 16, 261).

56 Ibid., III, 8, 203.

57 Ibid., I, 3, 29.

58 See, for example, ibid., II, 2, 113, 115; II, 4, 127.

59 Ibid., II, 8, 155. Although at II, 6, 139 he also says that both male and female palms bear fruit. For Theophrastus the date palm was an exotic, unusual and atypical tree.

60 Ibid., III, 9, 213, 215.

61 In *Enquiry into Plants* Theophrastus mentions, for example, differences between 'male' and 'female' in the silver fir (III, 9, 217), lime (III, 10, 225), maple (III, 11, 229), cornelian cherries (III, 12, 235) and Indian reeds (IV, 11, 379).

62 Lin Foxall ('Natural Sex: The Attribution of Sex and Gender to Plants in Ancient Greece', in Lin Foxhall and John Salmon, eds., *Thinking Men: Masculinity and Its Self-Representation in the Classical Tradition*, Routledge, London, 1998, 65), who has also noted 'the complexity and inconsistency of the use of the terms "male" and "female" for plants' in Theophrastus, characterizes the masculinity and femininity of plants as two kinds of nature needing to be tamed by ancient Greek man. Since all femininity was to be tamed, the qualities ascribed to female trees are not so far removed from the qualities ascribed to Greek women. The more complicated case of the wild masculine forms she explains as the projection of the masculinity of savages and foreigners, also needing to be conquered. Jennifer Robertson ('Sexy Rice: Plant Gender, Farm Manuals and Grass-Roots Nativism', *Monumenta Nipponica*, Vol. 39, No. 3 (Autumn 1984): 233–60) outlines the practice and systems of 'gender attribution' of rice in the Edo period. These are in some ways similar to Theophrastus's remarks on male and female in plants, to the extent that the gender attribution in question is explicitly distanced from sex attribution; that is, it is not about identifying male and female as 'partners' in reproduction. In the systems Robertson describes the female is in all cases, and perhaps even by definition, superior. Attribution was based on morphological aspects.

On Theophrastus see also Moshe Negbi, 'Male and Female in Theophrastus's Botanical Works', *Journal of the History of Biology*, Vol. 28, No. 2 (Summer 1995): 317–32; Marine Bretin-Chabrol et Claudine Leduc, 'La botanique antique et la problématique du genre', *Clio. Femmes, Genre, Histoire*, Vol. 29 (2009): 205–23.

63 See, for example, *Causes of Plants* (*De Causis Plantarum*, trans. Benedict Einarson and George K. K. Link, Heinemann, London /Harvard University Press, Cambridge MA, 1976), 45; 13, 1, 105.

64 Ibid., 16, 5–6, 133.

65 Ibid., 185, 279.

66 As we discuss in Chapter 4, Camerarius understood Theophrastus's characterization of male and female plants in this way. Theophrastus's first

century CE successor, Dioscorides, in his *De Materia Medica*, held much the same view. See John M. Riddle, *Dioscorides on Pharmacy and Medicine*, University of Texas Press, Austin, 1985, 31: 'Dioscorides occasionally ascribed sexual gender to plants but, as with all his fellow Greco-Romans, there was no order to it and usually no rationale. In general, male plants were considered harder, rougher, drier, and more barren, whereas female plants were softer, smoother, moister, and more fruitful. In the case of fig and date trees, the ancients correctly distinguished between male and female in the same way that modern botany does. Dioscorides employed genders sometimes to distinguish similar species of the same genus… As Theophrastus had done before him, he gave recognition to popular folk use of gender designations without the imposition of a theory or the study of a method.' See also R. J. Harvey-Gibson, *Outlines of the History of Botany*, A&C Black, London, 1919, 7: 'That plants as a group has sex like animals he [Theophrastus] did not realise, nor, for that matter, did any other botanist for two thousand years after his day. For although he talks of trees as "male" and "female", "male" stands to him as a synonym for "barren".' Also, J. Reynolds Green, *A History of Botany in the United Kingdom from the Earliest Times to the End of the 19th Century*, J. M. Dent, London & Toronto, 1914, 139. Francis Bacon held to the same view in 1639: 'the difference of *Sexes* in *Plants*, they are oftentimes by name distinguished; as *Male-Piony, Female-Piony; Male Rosemary, Female-Rosemary; He-Holly, She-Holly, &c*. But Generation by Copulation (certainly) extendeth not to Plants… Nevertheless, I am apt enough to think, that this same *Binarium* of a stronger and a weaker, like unto *Masculine* and, *feminine*, doth hold in all Living Bodies. It is confounded sometimes; as in some Creatures of Putrefaction, wherein no marks of distinction appear; and it is doubled sometimes, as in Hermaphrodites: but generally there is a degree of strength in most Species' (Francis Lord Verulam Viscount St Albans (Bacon), *Sylva Sylvarum, or a Naturall History in Ten Centuries*, ed. W. Rawley, William Lee, London, 1639, 126–7).

67 Aristotle's *History of Animals*, V 1, 539a21 mentions 'my treatise on *Plants*' but no such treatise survives. The Arabic translation of Nicolaus's *De Plantis* by Ishāq ibn Hunayn (c. 900 CE) attributes the work to Aristotle, as does Albertus Magnus in his commentary. Scaliger's 1566 commentary on *De Plantis* presents it as pseudo-Aristotelian, but it was still being attributed to Aristotle into the nineteenth century. It seems that Nicolaus's *De Plantis* was originally written in Greek. Fragments of a Syriac translation survive in writings by Bar Hebraeus. The full Syriac translation was translated into Arabic (by Ishāq ibn Ḥunayn and by Ibn At-Tayyib, with surviving manuscripts), and the Arabic translated into Latin (by Alfredus Anglicus) and Hebrew (by Shemtov ibn al-Falaquera and by Qalonymos), with an anonymous retroversion into Greek that was then translated into English by unknown translators. On the history of the text see H. J. Drossaart Luflos and E. L. J. Poortman, eds., *Nicolaus Damacenus, De Plantis: Five*

Translations, North Holland Publishing Company, Amsterdam, 1989, from where the above information is mostly drawn. *De Plantis*, in a modern English translation of Meyer's 1841 edition of Alfred Anglicus's twelfth-century CE Latin translation, appears in the Princeton edition of *The Complete Works of Aristotle*, Volume Two, marked as spurious. In the Loeb Classical library there is a translation of the Greek retroversion, said to be translated from an unspecified Medieval Latin edition. The various translations gathered by Drossaart Luflos and Poortman give us several texts that agree in much but not everything. But for simplicity's sake we refer here to 'the text' of Nicolaus' *De Plantis*.

68 Drossaart Luflos and Poortman, *Nicolaus Damacenus, De Plantis*, 56. See also 126, 138 (Ḥunayn); 450 (from the Hebrew 'Huntingdon' Fragment). Interestingly, the Latin translations introduce the word 'sexus' for male and female, which Drossaart and Poortman (250) describe as 'an example of the arbitrary handling of the text in Alfred's translation'. The evidence is from Aristotle, *Generation of Animals*, I xxiii, 731a1–14): 'In plants, however, these faculties [male and female] are mingled together; the female is not separate from the male; and that is why they generate out of themselves, and produce not semen [*gonēn*] but a fetation [*kuēma*] – what we call their "seeds" [*spermata*]. Empedocles puts this well in his poem, when he says "So the great trees lay eggs; the olives first… " because just as the egg is a fetation from part of which the creature is formed while the remainder is nourishment, so from part of the seed is formed the growing plant, while the remainder is nourishment for the shoot and the first root. And in a sort of way the same thing happens even in those animals where male and female are separate; for when they have need to generate they cease to be separate and are united as they are in plants' (The Fragment from Aristotle is known as 79 in Empedocles *Poem*).

69 Ḥunayn, 128, in Drossaart Luflos and Poortman, *Nicolaus Damacenus, De Plantis*.

70 For example, Ḥunayn 138, Ibn At-Tayyib 218, in ibid.

71 Ibn At-Tayyib 218, in ibid.

72 Bar Hebraeus, 74, in ibid.

73 Bar Hebraeus, 76, in ibid.

74 Ḥunayn 140, in ibid.

75 See also Ibn At-Tayyib, 218; Alfredus 523, in ibid.

76 Alberti Magni, *De Vegetablibus Libri VII*, Editionem Criticam ab Ernesto Meyero Coeptam, Berolini, 1867, 40–1. We gratefully use the unpublished translations of parts of *De Vegetabilibus* made for us by Dr Lee Raye.

77 Alberti Magni, *De Vegetablibus*, 41.

78 Jerry Stannard characterizes the distinction between male and female plants in Albertus (like the distinctions larger/smaller and summer/winter) as 'taxonomic'. Stannard, 'Albertus Magnus and Medieval Herbalism', in J. Stannard, K. E. Stannard and R. Kay, eds., *Pristina Medicamenta*, Ashgate, Aldershot, 1999, 367.

79 Alberti Magni, *De Vegetablibus*, 84–5, 87.

80 Alberti Magni, *De Vegetablibus*, 88: 'masculinum et femininum esse inseparabilia accidentia et propria animalium, et non plantarum'. On the idea of 'inseparable' or 'essential' accidents see Sandford, 'From Aristotle to Contemporary Biological Classification: What Kind of Category Is "Sex"?', 8–10.

81 Alberti Magni, *De Vegetablibus*, 89.

82 Ibid., 90, 91. The following from *Generation of Animals* is perhaps an example of Aristotle's 'metaphorical' statements: 'Indeed, animals seem to be just like divided plants: as though you were to pull a plant to pieces when it was bearing its seed and separate it into the male and female present in it' (*Generation of Animals* I, XXIII, 731a22–4). David Twetten and Steven Balder ('Introduction', in Irven M. Resnick, ed., *A Companion to Albert the Great: Theology, Philosophy and the Sciences*, Brill, Leiden, 2013, 168) and Karen Reeds ('Albert on the Natural Philosophy of Plant Life', in James A. Weisheipl, ed., *Albertus Magnus and the Sciences: Commemorative Essays, 1980*, Pontifical Institute of Medieval Studies, Toronto, 1980, 344–5) emphasize that Albertus is aiming not simply to give a correct account of or commentary on Aristotle's position but also to correct him and to provide 'his own formulation of what Aristotle had really meant to say about the natural philosophy of plants' (Reeds, 344) even if he did not actually say it.

83 Alberti Magni, *De Vegetablibus*, 92, 93. This conclusion on Albertus's position is also found in Stephan Fellner, *Albertus als Botaniker*, Wien, Alfred Hölder, Wien, 1881, 34, 62–5.

Chapter 3

1 Harvey-Gibson, *Outlines of the History of Botany*, 21. See also John W. Harshburger, 'James Logan, an Early Contributor to the Doctrine of Sex in Plants', *Botanical Gazette*, Vol. 19, No. 8 (August 1894), 307: only 'spasmodically, and often in an uncertain and indirect way did leaders in botanical thought break away from the scientific mysticism of the ancients. The path forward was a long and tortuous one'.

2 Theophrastus, *Enquiry into Plants*, I, iii, 22/3–24/5: 'A tree [*dendron*] is a thing which springs from the root with a single stem, having knots and several branches, and it cannot easily be uprooted; for instance, olive, fig, vine. A shrub [*thamnos*] is a thing which rises from the root with many branches; for instance, bramble, Christ's thorn. An under-shrub [*phuganon*] is a thing which rises from the root with many stems as well as many

branches; for instance, savory rue. A herb [*poa*] is a thing which comes up from the root with its leaves and has no main stem, and the seed is borne on the stem; for instance, corn and pot-herbs.' On the early history of herbals and the beginnings of plant classification see Agner Arber, *Herbals: Their Origin and Evolution*, Lost Library, Glastonbury, undated [1912]. She notes, for example (138), that Hieronymous Bock (aka Tragus) used Theophrastus's scheme in his *New Kreutterbuch* (1539).

3   See T. A. Sprague, 'Plant Morphology in Albertus Magnus', *Bulletin of Miscellaneous Information (Royal Botanic Gardens, Kew)*, No. 9 (1933): 431–40, 432; Arber, *Herbals*, 134–6.

4   Julius von Sachs (*History of Botany (1530–1860)*, trans. Henry E. F. Garnsey, Clarendon Press, Oxford, 1906 [1871–2], 40) calls Cesalpino's *De plantis* a work 'which forms an epoch in the science' of botany. See also Ellison Hawks, *Pioneers of Plant Study*, Books for Libraries Press, Freeport New York, 1928, 186: 'in its taxonomy [Cesalpino's] *De Plantis* most decidedly marked an epoch in the history of Botany.' See also Morton, *History of Botanical Science*, 137; Harvey-Gibson, *Outlines of the History of Botany*, 19–20. Linnaeus (*Philosophia Botanica*, trans. Stephen Freer, Oxford University Press, Oxford, 2003 [1751], 31) calls Cesalpino 'a fructist and the first true systematist'.

5   See footnote 28, below.

6   The *Naturphilosophie* of nineteenth-century German thought is an historical exception, but the traditional histories of botany see this as an aberration, not a scientific contribution.

7   After Albertus, the tradition of Aristotelian botany was for some centuries eclipsed by the development of medicinal herbals. As Arber explains (*Herbals*, 136), the herbalists' interest in individual plants and their medical properties required convenient ways of listing or arranging knowledge of a significant number of kinds of plants, but the demand of utility could be met (with, for example, an alphabetical arrangement) without any attempt to classify natural groups of plants – classification was not itself an intellectual concern. For the medicinal herbalists and those who would use their books, description and identification of specific plants for specific purposes was most important.

8   Andreae Cesalpini Arentini [Cesalpino], *De plantis libri XVI*, Apud Georgium Marescottum, Florentiae, 1583, 1; translated by Sachs, *History of Botany*, 43, 44.

9   Ibid., Ch XIII, 26; translation by Morton, *History of Botanical Science*, fn 46, 158. For Cesalpino on classification see also A. G. Morton, 'Marginalia to Cesalpino's Work on Botany', *Archives of Natural History*, Vol. 10, No. 1 (1981): 32; Brian W. Ogilvie, *The Science of Describing: Natural History in Renaissance Europe*, University of Chicago Press, Chicago IL, 2006, 224–5.

10 See C. E. B. Bremekamp, 'A Re-Examination of Cesalpino's Classification', *Acta botanica neerlandica*, Vol. 1, No. 4 (1953): 580–93, 586.

11 A. G. Morton, 'A Letter of Andrea Cesalpino', *Archives of Natural History*, Vol. 14, No. 2 (1987): 170, 171. As we have already noted, 'philosophy' and what we now call 'science' were not separate activities in the sixteenth century, so here by 'philosophy' Cesalpino means 'theoretical' work. *The* theoretical authority was Aristotle.

12 Book I of Cesalpino's *De plantis* is a discussion of the general nature of plants; the following fifteen Books give an account of the various plants known to Cesalpino, arranged according to the method of classification explained in Book I.

13 Cesalpino, *De plantis*, 2–3; translations from Sachs, *History of Botany*, 45–6. Cesalpino (3) identified the point at which the root joins the shoot as the most likely and suitable place for the location of the heart. On Cesalpino's use of analogies with animals see also Atran, *Cognitive Foundations of Natural History*, 225.

14 Cesalpino, *De plantis*, 11–12; translation from Sachs, *History of Botany*, 48.

15 Cesalpino *De plantis*, 15. See Maurice Dorolle, 'Introduction', in Césalpin, *Questions Péripatéticiennes*, trans. Maurice Dorolle, Librarie Félix Alcan, Paris, 1929, 78–9. It is presumably Cesalpino that Morton (*History of Botanical Science*, 213) has in mind when he writes that there was no revival of interest in the question of plant sexuality 'even in the sixteenth century revival of botany' with botanists 'content to repeat what Theophrastus had said and leave it at that'. But in fact Cesalpino broaches the philosophical issue of male and female in plants where Theophrastus does not.

16 Andreae Cesalpini Arentini, *Peripateticarum Quaestionem*, Venetiis, Venetiis, 1571, 97. On this point see Walter Pagel, 'Cesalpino's Peripatetic Questions', *History of Science*, Vol. 13, No. 2 (1973): 130–8, 133.

17 Holger Funk, 'Adam Zalužanský's "De sexu plantarum" (1592): An Early Pioneering Chapter on Plant Sexuality', *Archives of Natural History*, Vol. 40, No. 2 (2013): 244–56. This article includes Funk's translation of Zalužanský's chapter.

18 Zalužanský, 'De sexu plantarum', §3: 254; §6, 7: 257; §4: 254.

19 Sachs, *History of Botany*, 381.

20 Zalužanský, 'De sexu plantarum', §5: 254.

21 Sachs (*History of Botany*, 379–80) suggests that Cesalpino denies outright that plants have male and female, but this is because Sachs assumes that 'male' and 'female' can only mean male and female individuals. See also L. C. Miall, *The Early Naturalists: Their Lives and Work* (*1530–1789*), Macmillan & Co, New York, 1912: For Cesalpino 'Plants have no sexes, because in them the *genitura* is not distinct from the *materia*' (37).

22 For Morton (*History of Botanical Science*, 130) Cesalpino 'illustrates how old and new ideas continued to exist side by side in early Renaissance botany'.

23 Published by W. Rawlins.

24 In the first, 1673, edition of this text the title is *An Idea of a Phytological History of Plants*. It was published by Richard Chiswell, together with *A Continuation of the Anatomy of Vegetables, Particularly Prosecuted upon Roots. And an Account of the Vegetation of Roots Grounded Chiefly Thereupon*.

25 Grew, *The Anatomy of Plants. With an Idea of a Philosophical History of Plants, and Several Other Lectures Read before the Royal Society*, W. Rawlins, 1682, §22–23, 39; §25, 40.

26 Grew, *The Anatomy of Plants*, Chapter V, §4, 171–2. In fact, Millington was the Sedleian (not Savilian) professor of natural philosophy at Oxford (1675–1704).

27 Hawks, *Pioneers of Plant Study*, 220. As discussed below, Hawks considerably overstates what Grew and Ray recognized. On Grew's primary role in the discovery of sex in plants see also, for example, Samuel Morland, 'Some New Observations upon the Parts and the Use of the Flower in Plants', *Philosophical Transactions*, 1703, Vol. 23, No. 287 (1 January 1703): 1474–9, 1474; Patrick Blair, *Botanick Essays in Two Parts*, Printed by William and John Innys … at the Prince's Arms, London, 1720, Preface (unpaginated); Charles E. Raven, *John Ray: Naturalist. His Life and Works*, Cambridge University Press, Cambridge etc, 1950 (second edition) [1942], 11.

28 Taiz and Taiz, *Flora Unveiled*, 493. See also Agnes Arber, 'Nehemiah Grew 1641–1712', in F. W. Oliver, ed., *Makers of British Botany: A Collection of Biographies By Living Botanists*, Cambridge University Press, Cambridge, 1913, 61: 'Grew's most interesting contribution to science was, perhaps, his publication of the fact that the flowering plants, like animals, shew the phenomena of sex'. Harvey-Gibson (*Outlines of the History of Botany*, 43–4) says that Sachs and the 'continental historians' give the credit to Camerarius (*De sexu plantarum*, 1694), but Grew (and indeed Ray) had got there first: Camerarius's contribution is 'really an experimental proof of the correctness of the theory suggested by Grew and supported by Ray'. So too J. Reynolds Green, *A History of Botany*, 146: Contra Sachs, 'we may claim with Pulteney that the honour of the discovery "that the sexual process was universal in the vegetable kingdom, and that the dust of the antherae was endowed with an impregnating power" is due in the first place to an English botanist [Grew]'. John Ray himself, in his *Historia Plantarum Generalis* (Samuelis Smith, London, 1643, Vol. I, Book I, Chapter X, 17), attributes the discovery to Grew. See also Raven, *John Ray: Naturalist*, 11.

29 Grew, *Anatomy of Plants*, §5, 172.

30 Grew, *Cosmologia Sacra, or a Discourse of the Universe as It Is the Creature and Kingdom of God, etc.*, W. Rogers, S. Smith, and B. Walford, London, 1701. See also 'Concerning the Nature, Causes and Power of Mixture', 1674, in *Anatomy of Plants*.

31 'Concerning the Nature, Causes and Power of Mixture', Chapter II, 'Of the Principles of Bodies', §1–5, 223.

32 Ibid., §1–10, 225–6; §12–14, 227. This also means, according to Grew, that 'we easily understand, how divers of the same *Principles*, belonging both to *Vegetables* and many *other* Bodies, are also *actually* existent in the Body of *Man*. Because even in *Generation* or *Transmutation*, the *Principles* which are translated from one Body to another, as from a *Vegetable* to an *Animal*, are not in the least *alter'd* in themselves; but only their *Mixture*, that is, their *Conjugation*, *Proportion* and *Location*, is *varied*.' (Chapter III, 'Of the Nature of Mixture', §15, 227).

33 Grew, *The Anatomy of Plants*, Book IV, §5, 172. Malpighi was one of those who thought that the pollen was an excrescence of superfluous matter, comparing it to the menstrual blood of animals. In Ray's account of Malpighi (in Lazenby's translation) he says: 'Hence (says Malpighi), using a perhaps not incongruously derived name, menstrual purgations, which in women closely precede the times of conception, are called flowers. For as a fixed portion of the sap in plants is distilled through the stamens and petals of the flower, so in viviparous creatures, because these can in some way affect the particles of conception, they are sifted and thrust out each month through the uterus, so that the rest of the refined blood languishing in the uterus may more easily be made fertile by the power of the seed and may be channelled into the nature of the animal.' (In Ray, *Historia Plantarum Generalis*, trans. Margaret Lazenby, Vol. 3, 1072. [Newcastle University thesis, available online.]

34 Grew, *The Anatomy of Plants*, Book IV, §6, 8–10, 172–3.

35 Grew, *Anatomy of Plants*, Bk I, Ch. V, §17, 38.

36 Ibid., 38–9. Grew refers to tables IV and 59. Grew says that the blade commonly divides into two at the top and that there are commonly 'globulets' on its point.

37 Taiz and Taiz, *Flora Unveiled*, 246. See also, for example, Raven, *John Ray: Naturalist*, 45; Kassinger, *A Garden of Marvels*, 248. Commenting on Grew's illustrations Taiz and Taiz (*Flora Unveiled*, 333) claim that when Grew notes the phallic appearance of the attire (which, not quite accurately, they equate with the stamens; the attire includes the pistil, for Grew) his illustrations of the stamens 'had to greatly exaggerate the width of the filaments relative to the anther to achieve the effect'. (The illustration is on p. 334 of *Flora Unveiled*.) But Grew does not ever explicitly compare the stamen to the penis, and the table to which they refer – 56 – does not include an illustration of a blade. Elsewhere, (334) Taiz and Taiz admit: 'his phallic interpretation of its [the blade's] appearance is not only fanciful, it is based on an erroneous attribution of sex'.

38 As both Arber ('Nehemiah Grew 1641–1712', 61) and Sachs (*History of Botany*, 383) do note.

39 Grew, *Anatomy of Plants*, 37–8. In the more detailed description in the paper on the anatomy of flowers he describes the seminiform attire as a little sheaf of seed-like particles, 'standing on so many *Pedicills*, as the *Ear* doth upon the *End* of the *Straw*.' Chapter III, §2, p. 167. John Ray also used the Latin *pediculus*; what in modern botany is called the 'filament' of the stamen. See, for example, Ray, *Historia Plantarum*, Book I, Chapter X, 16–17.

40 Grew, *Anatomy of Plants*, Chapter III, §9, 168.

41 For example, as Arber ('Nehemiah Grew 1641–1712', 61) says, Grew 'falls into the error of supposing that the pollen grains are in some cases originally produced by the style and stigmas, which he calls the "Blade," and which he did not recognise as part of the female organ. His figures make it clear that he mistook the stylar hairs for little stalks organically connecting the pollen grains and the style'.

42 Grew, *Cosmologia Sacra*, §1, 31.

43 Ibid., §8–9, 32. Grew argues (§11, 32–3) that the corporeality of all bodies is the same and that the degrees of subtility of corporeality is also the same in all bodies. If it were the case, then, that subtility produced life, then any degree of subtility would produce life in any body, and a greater degree of subtility would produce a greater degree of life – all of which, he says, is 'Subtile Nonsense'.

44 Ibid., §12–13, 33. Grew also says that if organization was the condition of life, it would also be the case that a more organized or complex body would be more alive – that is, that degree of complexity of organization would correlate with degrees of vitality (§13, 33).

45 Ibid., §16, 33. Grew does not distinguish here between a body which is moved and one which moves itself, but he had effectively dealt with this in the first Book. There he argued that motion is not of the essence of body, as we can conceive of body without motion, and thus (as far as he is concerned) motion must be 'something else besides Body'. Because motion is something other than body, if body could move itself it would be as if it had 'the Power of Making Something of Nothing', which would be akin to its having the 'Power of Being' or being 'Self-existent'. God, of course, is the creator of both matter and motion (Book I, Chapter I, §32–34, 4–5). In Book II (§16, 33) he also claims, on the back of this, that all motion (whether in living bodies or human made artefacts) is essentially the same.

46 Ibid., §19, 34. Grew means 'subject' in the ontological sense here – the underlying substance.

47 Ibid., §18, 34.

48 Ibid., § 20, 34.

49 Ibid., Book I, Chapter I, §32, 4.

50 Ibid., § 29, 35.

51 Ibid., §30, 35.

52 Ibid., §24, 34.

53 Ibid., § 33–36: 'There are Sundry Motions, both in Plants and Animals, depending upon this Vegetable Life' (§35, 35). In animals (including humans) these include the motions of the guts and the heart, and all those operations which continue while we are asleep and of which we have no perception (§34–35, 36).

54 See ibid., Book II, Chapter II, 'Of Sense', and Chapter III and IV on mind. 'Nor do I see any reason at all, why the Vital Principles of Things, as well as the Corporeal, may not be compounded. Provided, that as the Mixture of the Corporeal, is suitable to the Nature of every Part: So the Union of the Vital, to that of the whole' (§2, 37).

55 Grew, *Cosmologia Sacra*, Chapter I, §31, 35. As Brian Garrett notes, Grew is thus unusual in being committed, simultaneously, to atomistic and mechanistic explanations with regard to matter *and* to a vital principle (Brian Garrett, 'Vitalism and Teleology in the Natural Philosophy of Nehemiah Grew (1641–1712)', *BJHS*, Vol. 36, No. 1 (March 2003): 63–81, 65). A similar explanatory bifurcation (mechanism in the laws of nature; teleology in life) characterizes Kant's philosophy at the end of the eighteenth century.

56 Grew, *The Anatomy of Plants*, Book IV, §6, 8–10, 173.

57 Ray, *Historia Plantarum Generalis*, Vol. I, Book I, Chapter X, 17: '*sed solus ejus halitus effluvia subtilia sufficiunt ad ova foecundanda, & embryon intus conclusum vivificandum*'. Raven, for example, glosses Ray's sentences on Grew as referring to his claim that the stamens are male sex organs, an observation supported by reference to dioecious plants; no mention of the vivifying 'office' (Raven, *John Ray*, 223). Ray also repeats Grew's claim that it is not incredible that plants should be both male and female at the same time, as some animals are.

58 On this issue in Ray's *Stirpium Europaearum* see Raven, *John Ray*, 285. To the extent that Grew identifies the vivifick principle with the vegetative soul, Ray disagrees with him. Like Grew, Ray was convinced that matter itself was not alive, and like Grew he expounded on this in a philosophical and religious work, *The Wisdom of God, Manifested in the Works of Creation*, Facsimiliae of 1826 edition, Scion Publishing, Bloxham, Oxon, 2005: 'Let Matter be divided into the subtilest Parts imaginable, and these be mov'd as swiftly as you will, it is but a senseless and stupid Being still, and makes no nearer Approach to Sense, Perception or vital Energy, than it had before; and do but only stop the internal Motion of its Parts, and reduce them to Rest, the finest and most subtile Body that is, may become as gross, and heavy, and stiff as Steel or Stone.

'And as for any external Laws or established Rules of Motion, the stupid Matter is not capable of observing or taking any Notice of them, but would be as sullen as the Mountain was that Mahomet commanded to come down to him; neither can those Laws execute themselves. Therefore there must, besides Matter and Law, be some Efficient, and that either a

Quality or Power inherent in the Matter itself, which is hard to conceive, or some external intelligent Agent, either God himself immediately, or some Plastick Nature' (*Wisdom of God*: 49–50). Ray follows Ralph Cudworth in believing that this power is not immediately God himself but 'a Plastick Nature; for the Reasons alledg'd by Dr. Cudworth, in his System' (51). Ray is unwilling to attribute this power to the vegetative soul because the fact that plants grow from cuttings means that the vegetative soul is divisible, 'and consequently no spiritual or intelligent Being; which the plastick Principle must be, as we have shewn: For that must preside over the whole Oeconomy of the Plant, and be one single agent, which takes Care of the Bulk and Figure of the whole, and the Situation, Figure, Texture of all the Parts, Root, Stalk, Branches, Leaves, Flowers, Fruit, and all their Vessels and juices. I therefore incline to Dr. Cudworth's Opinion, that God uses for these Effects the subordinate Ministry of some inferiour Plastick Nature; as in his Works of Providence he doth of Angels' (52). Concerning plants in particular, how do we explain that plants preserve their kind even when they are transplanted to different soils, if not by 'some *logos spermatikos*, Seminal form or virtue which doth effect this or rather some intelligent plastick Nature; as we have before intimated: For what Account can be given of the Determination of the Growth and Magnitude of Plants from Mechanical Principles, of Matter mov'd without the Presidency and Guidance of some superior Agent?' (101).

59 Conway Zirkle, 'Introduction' to N Grew, in *The Anatomy of Plants*, Johnson Reprint Corp., New York and London, 1965, xiv.

60 Hawks, *Pioneers of Plant Study*, 211.

61 Sachs, especially, tends to see Grew's philosophy in this light (*History of Botany*, 239): 'while in Malpighi we only occasionally encounter the philosophical prejudices of his time, which usually lead him into mistakes, Grew's treatise is everywhere interwoven with the philosophical and theological notions of the England of that day; but we are compensated for this by the more systematic way in which he pursues the train of thought, and especially by the constant effort to give as clear a representation as possible of what he sees'. Sachs dismisses the discussion of the vivifick effluvia etc as 'some curious remarks' (383).

Chapter 4

1 Sachs, *History of Botany*, 387.

2 Ibid., 389.

3 The idea that Camerarius refers to Aristotle out of a sense of duty sits uneasily with the fact that Camerarius seemed to feel the need to justify his appeals to Aristotle and with the extent of the reference to Aristotle.

Whole pages are devoted to Aristotle and various Aristotelians, and an Appendix in the first, 1694, publication of the letter includes commentary on passages from *De Plantis* (which Camerarius attributes to Aristotle) and from *Generation of Animals*. Rudolphi Jacobi Camerarii [Camerarius], *De sexu plantarum epistola*, Tubingae, 1694. On his use of Aristotle see, for example, 118.

4   Even an empiricist like John Stuart Mill was happy to accept a good analogical argument, 'inferring one resemblance from other resemblances without any antecedent evidence of a connection between them… may approach in strength very near to a valid induction' (Mill, *A System of Logic, Ratiocinative and Inductive*, Harper & Brothers, New York, 1882, 688). On the use of analogy and analogical models in the sciences see, for example, Harry F. Olson, *Dynamical Analogies*, Chapman & Hall, London, 1943; Max Black, *Models and Metaphors: Studies in Language and Philosophy*, Cornell University Press, Ithaca NY, 1962; W. H. Leatherdale, *The Role of Analogy, Model and Metaphor in Science*, North-Holland Publishing Company, Amsterdam/Oxford, 1974; Colin Murray Turbayne, *The Myth of Metaphor*, University of South Carolina Press, Columbia SC, 1970; Fernand Hallyn, ed., *Metaphor and Analogy in the Sciences*, Kluwer, Dordrecht, 2000.

5   See Nancy Leys Stepan, 'Race and Gender: The Role of Analogy in Science', *Isis*, Vol. 77, No. 2 (June 1986): 271.

6   Black, 'Models and Archetypes', in his *Models and Metaphors*, 242. See also Brian Vickers, 'Analogy Versus Identity: The Rejection of the Occult Symbolism 1580–1680' in Vickers, ed., *Occult and Scientific Mentalities in the Renaissance*, Cambridge University Press, Cambridge, 1984. Occult tradition (unlike scientific mentality) collapses analogy into identity: 'One could sum up the difference between the occult and the experimental scientific traditions at this point in the form that where the experimentalist will say "this is not reality, but only a trope," the occultist will say "this is not just a trope, but reality"' (136). This is the main danger against which we are counselled to guard in Turbayne's *The Myth of Metaphor* (see, for example, p. 3). As John Tyler Bonner points out ('Analogies in Biology', *Synthese*, Vol. 15 (1963): 275), there is also the danger that we forget that what is alike in some respects is not alike in all respects. Effectively there is the danger that, in relation to the same things being compared, we construct further illegitimate analogies on the basis of a legitimate analogy. This is a species of a problem identified by Olson: 'The temptation to ignore negative analogies in our pleasure with seeing positive analogies is understandable and compelling' (Introduction, *Science as Metaphor*, 6). M. Barker ('Putting Thought in Accordance with Things: The Demise of Animal-Based Analogies for Plant Functions', *Science & Education*, Vol. 11, No. 3 (2002): 293–304) is interesting on the confusions that the use of analogy may introduce into the teaching of plant biology: 'On the one hand, when the topic is "Living Things", we

teachers focus on the common features (respiration, excretion, sexual and asexual life cycles) and we would encourage learners to transfer their knowledge of animal functions by analogy to plants (and vice versa). In another topic area, "Differences Between Plants and Animals", we emphasise the unique features (autotrophic versus heterotrophic nutrition; transpiration versus blood circulation) and we somehow expect our students to know that analogies are now not appropriate' (302). Concerning the sexuality of plants, though, Barker sees only a history of negative analogy: 'In terms of sexual reproduction, the analogy worked in a negative sense (analogists held that the immobile plants did not reproduce sexually); this view persisted until experimentalists gradually established the pervasiveness of gamete fusion in plants in the eighteenth and nineteenth centuries' (295). Barker is wrong on this last point: acceptance of plant sexuality predated any knowledge of gamete fusion or indeed of gametes themselves. We consider the implications of this at the end of this chapter.

7 Arber, 'Analogy in the History of Science', 232–3. Georges Canguilhem suggests that the longevity of analogy in biology can be explained with reference to the central importance of the concept of the 'organ' (Canguilhem, 'Modèles et analogies dans la découverte en biologie', in *Études d'histoire et de philosophie des sciences*, Vrin, Paris, 1970 [1968], 305–6). Yves Gingras and Alexandre Guay ('The Uses of Analogies in Seventeenth and Eighteenth Century Science', *Perspectives on Science*, Vol. 19, No. 2 (2011)): 154–91 provide a comprehensive survey of the use of analogy in contributions to the *Philosophical Transactions of the Royal Society*, 1665–1780, and also note that if we apply modern disciplinary distinctions, analogy is most frequently used, by far, in biology and the medical sciences (160).

8 Arber, 'Analogy in the History of Science', 231. Arber refers to Grew's *The Anatomy of Plants*, 1682, which we discussed in Chapter 3.

9 Camerarius, *De Sexu Plantarum Epistola* (*Letter on the Sex of Plants*), in Johann Georg Gmelin, *Sermo Academicus de Novorum Vegetabilium Post Creationem Divinam Exortu*, Literis Erharddtianus, Tubinngae, 1749, 101–2; *Das Geschlecht der Pflanzen*, trans. M. Möbius, Wilhelm Engel, Leipzig, 1899, 15. References to Camerarius's text henceforth cite this Latin edition and then the German pagination.

10 Camerarius, *De Sexu Plantarum*, 113/25.

11 Gmelin, 'Praefatio' to *De Sexu Plantarum Epistola*, 62.

12 Camerarius, *De Sexu Plantarum*, 113–15/25–6.

13 Sachs, *History of Botany*, 389.

14 Camerarius, *De Sexu Plantarum*, 118–20/29–30.

15 Ibid., 122–3/32–3

16 Ibid., 123/32–3, 124–5/33–4, 125/34.

17 It is ultimately philosophical because it rests on the philosophical distinction between body (or matter) and spirit. This distinction played an important part in many seventeenth- and eighteenth-century theories of the physical world and in alchemical theories. But it is presupposed in them; it is not a result of them.

18 See, for example, Abbé Spallanzani, *Dissertations Relative to the Natural History of Animals and Vegetables*, [translator not credited], J. Murray, London, 1784: 'the embryo does not at all depend for its existence upon the powder of the stamina; therefore, secondly, the embryo exists in the ovarium independently of this powder; thirdly, nor is it the result of two principles, one depending on the pollen, and the other upon the pistil, as others suppose' (Unpaginated summary in index). Charles Alston, in his *A Dissertation on Botany,* Benjamin Dodds, London, 1754, also criticizes the 'sexualist' position at length.

19 Camerarius notes (*De sexu plantarum*, 118–19/28–9) that Aristotle distinguishes carefully between the semen of the male animal (*gonē*) and the plant seed (*sperma*), the latter being the result of the mixture of male and female.

20 Camerarius, *De sexu plantarum*, 117/28; the translation of this passage is taken from Taiz and Taiz, *Flora Unveiled*, 344.

21 See, for example, Sachs, *History of Botany*, 388.

22 Camerarius, *De sexu plantarum*, 115/26.

23 Julius von Sachs, *Lehrbuch der Botanik, Nach dem Gegenwärtigen Stand der Wissenschaft*, Wilhelm Engelmann, Leipzig, 1874, fourth edition, 870. Translation from Ulrich Kutschera and Karl J. Niklas, 'Julius Sachs (1868): The Father of Plant Physiology', *American Journal of Botany*, Vol. 105, No. 4: 656–66, 662.

24 Kutschera and Niklas, 'Julius Sachs', 662.

25 Lincoln Taiz, Eduardo Zeigler, Ian Max Møller and Angus Murphy, *Plant Physiology and Development*, Sinauer Associates and Oxford University Press, New York and Oxford, 2018, sixth edition, 625. See also Taiz and Taiz, *Flora Unveiled*, 493.

26 Mauseth, *Botany*, 206.

27 Claude Joseph Geoffroy, 'Observations sur la structure et l'usage des principals parties des fleurs', *Mémoires de mathématiques et de physique de l'Académie royale des sciences* (14 November 1711): 207–31, 217. On the experiments with corn see p. 223. But there is some doubt as to whether Geoffroy really did perfom these experiments himself or rather just repeated Camerarius's reports. See Paul Bernasconi and Lincoln Taiz, 'Sebastien Vaillant's 1717 Lecture on the Structure and Function of Flowers', *HUNTIA: A Journal of Botanical History*, Vol. 11, No. 2 (2002): 97–128, 101–3.

28 Richard Bradley, *New Improvements of Planting and Gardening, Both Philosophical and Practical, Explaining the Motion of the Sap and Generation of Plants &...*, W Mears, London, 1719 [1717], 20–4. The first parallel drawn between animals and plants, in order better to understand the latter, aims to show that sap circulates in the vessels of plants just as it does in the bodies of animals (4–5).

29 James Logan, 'Some Experiments Concerning the Impregnation of the Seeds of Plants,' *Philosophical Transactions*, Vol. 39, No. 4 (31 December 1735): 192–5. His letter begins: 'As the Notion of a male Seed, of the *Farina Foecundans* in Vegetables in now very common, I shall not trouble you with any Observations concerning it, but such as may have some Tendency to what I have to mention'; that is, proof that the male seed is necessary for the fertilization of maize. (192)

30 Morland, 'Some New Observations upon the Parts and Use of the Flower in Plants,' 1475. This was also Geoffroy's view, or at least that conjecture concerning generation in plants that he believed to be best founded; see Geoffroy, 'Observations sur la structure et l'usage des principales parties des fleurs,' 225. There is a sense in which the pollen is, indeed, a 'congeries of small plants,' as we shall see in Chapter 5 – but not as Morland imagined it.

31 Blair, *Botanick Essays*, unpaginated preface. Blair quotes Grew at length and concludes that things are the same in plants as in animals. As the seed is made up of the more gross and 'terrestrious' part, this shows the necessity for 'a more subtile, active Principle, to quicken, enliven and dispose this gross Substance of the Seed to fertility'; 'this Impregnation can be no otherways perform'd than by some *Emanation*, some *vivifick Effluvia*, some *prolifick Virtue* communicated by means of this *Farina*, or some other *Menstruum*, from the *Male* to the *Female* Parts of the *Plant*, by virtue of which the Parts of the *Seed* are dispos'd to be dilated, the *Tubuli Nutriviti* enlarg'd, a greater Supply of *Nourishment* to be furnish'd, and all the Particles composing the Seed so to be set in Motion and regulated, that they can be augmented, extended and increas'd to a due Proportion, which one Grain of small Dust, so confin'd could never do' (263; 298–9).

32 See Sachs, *History of Botany*, 390–402; Ellison Hawks, *Pioneers of Plant Study*, 221. Significant steps were also taken by Joseph Gottlieb Kölreuter, Karl Friedrich Gärtner and Giovanni Battista Amici.

33 Vaillant had been trained by Joseph Pitton de Tournefort, professor of botany at the Royal Gardens, who had denied the sexuality of plants until his death in 1708. Tournefort did, however, compare the seed to the egg, as in the Aristotelian botanical tradition. See Joseph Pitton de Tournefort, *Elemens de Botanique, ou Methode Pour Connoître les Plantes*, L'Imprimerie Royale, Paris, 1694, Tome I, 23.

34 Sébastien Vaillant, *Discours sur la structure des fleurs, leurs différences et l'usage de leurs parties*, Pierre Vander, Leiden, 1718, 16, 20; 'Lecture on the Structure and Function of Flowers,' 108, 109. The English translation is part of Bernasconi and Taiz's 'Sebastien Vaillant's 1717 Lecture'. References cite the French and then the English translation, but the translation is often modified.

35 Vaillant, *Discours*, 2/105, 10/107.

36 See, for example, Bernasconi and Taiz, 'Sebastien Vaillant's 1717 Lecture', 107.

37 Vaillant, *Discours*, 10/107.

38 Ibid.

39 Ibid., 6/106.

40 Ibid.

41 Ibid., 8/106.

42 Ibid., 14/108. He later (28/111) says that the ovary is the same things as the embryo (*embrion*) of the fruit.

43 Ibid., 16/108. Vaillant argues (20/109) that this volatile spirit, 'or, if I may dare to use the term from Genesis, this *breath*', can enter the belly of the ovary via the vessels in the trunk (style). When he moves on from the characterization of the sexual organs and the controversies over fertilization, Vaillant's lecture becomes more sober. He discusses whether and in which cases the calyx is a single or multifaceted part and the relation in different flowers between the number of stamens and the number of petals, finishing with a presentation of the arrangement of several new genera of plants.

44 Sachs, *History of Botany*, 398. John Ramsbottom ('S. Vaillant's Discours sur la structure des fleurs', *Soc. Bibiogr. Nat. Hist*, Vol. 4 (1718): 194–6) is among those who note, however, that Vaillant's proposal of the sexual parts of plants, and specifically the number of stamens, as a means of classification is important, not least because of the influence of this idea on Linnaeus.

45 In *Prelude to the Betrothal of Plants* (trans. Clare James, Uppsala Universiteitsbibliotek, Uppsala, 2007, 71) Linnaeus refers to 'the excellent Vaillant' and his *Discourse*, 'which I have not yet seen'. Linnaeus's diary is written in the third person: 'Linnaeus had, about this time [1729], read in the *Leipsic Commentaries*, a review of *Vaillant*'s treatise on the sexes of plants, by which his curiosity was excited to a close investigation of the *stamina* and *pistilla*; he observed that these parts of the plants were of essential importance, and that they varied as much as the petals. Hence he formed the design of constituting a new sexual method. There was just then published a philological dissertation *de Nuptiis Plantarum*, from the pen of *George Wallin*, librarian of the University; and as Linnaeus had no opportunity of publicly opposing it, or of starting his doubts, he drew up in writing a little treatise on the sexes of plants, in conformance with genuine botanical principles'. 'Linnaeus' Diary', in Richard Pulteney, *A General View of the Writings of Linnaeus, to Which Is Annexed the Diary of Linnaeus, Written by Himself*, J. Mawman, London, 1805, 519. It also seems that Johann Stensson Rothmann, one of Linnaeus's teachers at his school

in Växjö, introduced him to Vaillant. See, for example, Wilfrid Blunt, *The Compleat Naturalist: A Life of Linnaeus*, Frances Lincoln, London 2001 [1971], 17, 31; Gunnar Erikson, 'Linnaeus the Botanist', in Tore Frängsmyr, ed., *Linnaeus: The Man and His Work*, University of California Press, Los Angeles/London, 1983, 64; Knut Hagberg, *Carl Linnaeus*, trans. Alan Barr, Jonathan Cape, London, 1952, 66–7.

46 He notes that these are Vaillant's words for the pistil, stamens and anthers (*Prelude*, 80) and adopts them throughout (see, for example, 81, 85, 86, 88, 90).

47 The copy of Camerarius' *De sexu plantarum* in Linnaeus' library (now at the Linnean Society Library in London) comprises pages extracted from a larger book and bound into another. Its first page calls it the Appendix II of Michaelis Berhhardi Valentini Responsoria, dated 1701. It is inscribed on the inside front page in Linnaeus's own hand. James Edward Smith (who bought and first catalogued Linnaeus's library) dates it 1704. Although the handwriting is that of the young Linnaeus, it is not possible to say when he first saw it. A copy of the 1728 Latin edition of Vaillant's *Discourse* is also in the library at the Linnean Society but has no mark in Linnaeus's hand.

48 Linnaeus, *Prelude*, 90.

49 Ibid., 65.

50 That is, the view, like that of Theophrastus, that ascribed male and female to plants on the basis of 'the thickness or slenderness of the stalk… etc. etc'. Ibid., 68.

51 Ibid., 69–70

52 Ibid., 71.

53 Ibid., 72. If we reorder the parts of Linnaeus's argument the logic (such as it is) becomes clearer: as the flower is to the fruit what the animal male and female organs of generation are to the embryo (to the extent that both are, respectively, a precondition for fruit/embryo), the plant's organs of generation must be in the flower.

54 Ibid., 80. Later (Ibid., 91) Linnaeus justifies his analogy between plant seeds and animal ova. He writes (92): 'I call these seeds "ova", for the great Harvey called them by the same name 80 years ago when he refuted spontaneous generation and proclaimed the principle of *omnia ex ovo* to the world. The ova of plants are hatched in the soil, as those of quadrupeds hatch in the uterus and those of fish in water.' The two cotyledons of the germinating seed are then said to be analogous to the animal placenta (93).

55 Ibid., 94. 'Vagina' is probably intended here in its primary meaning of 'sheath'; it was only in the late seventeenth century that it came to refer specifically to the female genital passage. But, like Vaillant's use of 'queüe', the *double entendre* is probably meant. 'Vulva' is less ambiguous. It is perhaps Linnaeus that Tompkins and Bird (*The Secret Life of Plants*, 105–6) have in mind when they lament the loss of a straightforward vocabulary to describe plant sex: 'that plants have female organs in the form of vulva, vagina, uterus, and ovaries, serving precisely the same functions as they do in woman, as well as distinct male organs in the form of penis, glans, and testes, designed to sprinkle the air with billions of spermatozoa, were facts quickly covered by the eighteenth-century establishment with an almost impenetrable veil of Latin nomenclature, which stigmatized the labiate vulva, and mis-styled vagina; the former being called "stigma", the latter "style". Penis and glans were equally disfigured into "filament" and "anther"'.

56 Ibid., 94. Sten Lindroth ('The Two Faces of Linnaeus', in Frängsmyr, ed., *Linnaeus: The Man and His Work*, 10) notes, 'How close [Linnaeus] stands to the tradition of wedding poetry in the admired opening to the dissertation on the nuptials of flowers'. The word 'thalamus' is still used in botany to denote the structure at the tip of the pedicel out of which the floral parts arise. It has also been used to name a part of the brain since Galen (see Carlo Serra et al., 'Historical Controversies About the Thalamus: From Etymology to Function', *Journal of Neurosurgery* (Neurosurg Focus), Vol. 47, No. 3: E13, 2019.

57 Linnaeus, *Prelude*, 81.

58 Ibid., 81. Linnaeus was also writing in response to a public thesis presentation and disputation, which he had apparently not been able to attend, 'Concerning the Nuptials of Trees'. See Linnaeus's 'Diary' in Pulteney, *A General View of the Writings of Linnaeus*, 519; Blunt, *The Compleat Naturalist*, 31.

59 On the importance of Linnaeus's use of analogy throughout his work see James L. Larson, 'Linnean Analogy', *Scandanavian Studies*, Vol. 40, No. 4 (November 1968): 294–302.

60 The key is the 'Clavis Systematis Sexualis', in Caroli Linnaei, [Linnaeus], *Systema Naturae*, Theodorum Haak, Lugduni Batavorum [Leiden], 1735, unpaginated; Linnaeus, *A System of Vegetables*, trans. Erasmus Darwin and others, from the 13th edition of the Systema Vegetabilium… By a Botanical Society at Lichfield, Lichfield 1782/83. English translations are also taken from James Lee, *An Introduction to Botany Containing Explanations of the Theory of that Science and an Interpretation of Its Technical Terms, Extracted from the Works of Dr Linnaeus….*, Printed for J. and R. Tonson, in the Strand, London, 1760, 63–76.

61 In his translation of the 'Clavis' Robert John Thornton translates 'Nuptiae Publicae' as 'With the Sexes Visible' (Robert John Thornton, *A New Illustration of the Sexual System of Linnaeus*, Vol. 1, Printed for the author by T Bensley, London, 1799, 133).

62 Linnaeus, *Fundamenta Botanica*, Salomonem Schouten, Amstelodami, 1736, §146, 17.

63 Ibid., §147, 17.

64 Frans A. Stafleu, *Linnaeus and the Linnaeans: The Spreading of Their Ideas in Systematic Botany, 1735–89*, A Oosthoeks's Uitgeversmaatschappij, Utrecht, 1971, 26.

65 Linnaeus, *Philosophia Botanica*, §133, 100–1; Caroli Linnaei, *Philosophia botanica*, Godofr. Kiesewetter, Stockholm, 1751, 86–7. Subsequent page references are to the English and Latin, respectively.

66 *Philosophia Botanica*, §136, 102/89: 'The *cotyledons* of animals are produced from the yolk of the egg, in which the point of life is inherent; therefore the seminal leaves of the plants, which envelop the corcule, are the same.'

67 Ibid., §140–2, 102–3/96.

68 Ibid., §143–5, 105/96–7.

69 Ibid., §146, 105/92.

70 The 1760 *Dissertation on the Sexes of Plants,* trans. James Edward Smith, George Nicol, London, 1786, is an altogether more sober affair, as befits a public dissertation defended (if not authored) by one of Linnaeus's students. These dissertations are collected in a series of volumes entitled *Amoenitates Academicae* (meaning, roughly, 'academic pleasantries'). *Disquisitio de Sexu Plantarum* is found in Vol. 10, Dissertationes Botanicae, 1790, 100–31.

71 Linnaeus, *Praelectiones in Ordines Naturales Plantarum*, ed. Paulus Diet, Giseke, Impr, Benj. Gottl. Hoffmanni, Hamburgi, 1792, 16–18; translation from James Larson, 'Linnaeus and the Natural Method', *Isis*, Vol. 58, No. 3 (Autumn 1967): 304–20, 317. These lectures were published posthumously based on lecture notes from Giseke (the editor) and Fabricius. See Larson, 'Linnaeus and the Natural Method', 313. See also Linnaeus, *Dissertation on the Sexes of Plants*, 54, 55, 56.

72 See Larson, 'Linnaeus and the Natural Method', 315–16. The problem is noted by Linnaeus in *Philosophia Botanica*, § 175, 136/123.

73 See, for example, Antoine de Jussieu, *Du rapport des plantes avec les animaux tiré de la différence de leurs sexes*, 1721 Ms. 1260, Bibliothèque du Museum, reprinted as the Appendix to Delaporte, *Le second règne de la nature*. The English translation by Richard Bradley, 'The Analogy Between Plants and Animals, drawn from the Differences of their Sexes', in his *A Philosophical Account of the Works of Nature*, Printed for W. Mears, London, 1721, 25–48, is reprinted in the English translation of Delaporte's book, *Nature's Second Kingdom* (193–8). Jussieu begins with a series of analogies between plants and animals before touching on the sexual analogy; however, he did not think that the sexual parts of plants could rightfully be used as a means of classification (just as one does not classify animals according to their sexual organs). See also Philip Miller's entry on 'Generation' in his *The Gardener's Dictionary*, London, 1759 [1731]: 'The Generation of Plants bears a close Analogy to that of some Animals, especially such as want local Movement, as Mussels, and other immovable Shell Fish, which are Hermaphrodite, and contain both the male and the female Organs of Generation. The Flower of a Plant is found to be the *Pudendum,* or principal Organ of Generation.' Miller notes the two competing theories of plant generation at this time, one claiming that the plant is contained in miniature in the seed and needs only to be excited to fermentation by the 'farina', and the other claiming, like Morland, that the germ of the new plant is in the farina and only needs to find a suitable 'Nidus'. These theories, he says, 'bear a strict Analogy to those of animal Generation' (animaculism and ovism). Miller makes the argument for the sexes of plants with a point-by-point analogy with animals. Robert Hooper, in his *Observations on the Structure and Economy of Plants, to Which Is Added the Analogy between the Animal and the Vegetable Kingdom,* Oxford, 1797, follows Linnaeus's 'beautiful SEXUAL SYSTEM of vegetables' (10–11), describes 'the animal functions in plants' (automatic motion – that is, irritability – sleep and 'watching' or waking) (73–87), and the 'connubium or marriage of plants' and their 'parturition' (89–102), ending with an eighteen-page point-by-point explanation 'Of the Analogy between the Animal and the Vegetable' (109–28). The appeal to analogies with animals was still being made well into the nineteenth century, as in W. McGillivray, *Lives of Eminent Zoologists from Aristotle to Linnaeus,* Oliver & Boyd, London, 1834, 295: Linnaeus 'made the stamina and pistils the basis of his arrangement, which he was induced to do from the consideration of their great importance, as the parts most essential to fructification. These organs being analogous to those distinguishing the sexes of animals, the Linnaean method is sometimes called the sexual system'. Francis Darwin published a two-part essay on 'The Analogies of Plant and Animal Life' in *Nature* in 1878 (Pt I, 14 March; Part II, 21 March). On the continued use of animal-plant analogies in eighteenth-century science, and for more examples, see Philip C. Ritterbush, *Overtures to Biology: The Speculations of Eighteenth-Century Naturalists,* Yale University Press, New Haven & London, 1964 Chapters 3 and 4.

74 Miller, *Gardener's Dictionary,* in the entry under 'Generation'. See also Lee, *An Introduction to Botany,* vi. In Linnaeus's *Dissertation on the Sexes of Plants* (26) the phrase 'a real distinction of sex' (*sexus vera differentia*) refers specifically to dioecious plants ('the different sexes are in distinct individuals').

75 Julian Offray de la Mettrie, *L'Homme plante,* in *Oeuvres philosophiques,* Tome Second, Charles Tutot, 1796 [1747], 50.

76 Ibid., 54.

77 Ibid., 55.

78 Erasmus Darwin, The Botanic Garden. A Poem, Part II, in *The Loves of The Plants,* J. Johnson, London, 1799 [1789], 3.

79 Ibid., 5–6. Genista represents Decandria Monogynia – ten stamens, one pistil; Melissa represents Didynamia Monogynia – four stamens, two long (the 'knights') two shorter (the 'squires'), one pistil. On Darwin's poem see

Janet Browne, 'Botany for Gentlemen: Erasmus Darwin and "The Loves of the Plants"', *Isis*, Vol. 80, No. 4 (1989): 593–621. Meeker and Szabari (*Radical Botany*, 90) remark on the heteronormative aspects of Darwin's 'libidinal and libertine vegetality'. They also, however, interpret other speculative plant fictions based on analogies between plant and animals more positively. In these speculative fictions, they claim, these analogies 'are mobilized to call into to question the seemingly self-evident hierarchical relationship between humans and plants' (57). Perhaps their most persuasive reading concerns Ludvig Holberg's *The Journey of Niels Kim to the World Underground* (James I. McNeils Jr., ed., University of Nebraska Press, Lincoln, NE, 1960), where the physiology, social forms and mores of a planet of rational, intelligent trees provide a positive contrast with aspects of human existence and social life. See Meeker and Szabari, *Radical Botany*, 63–74.

80 Londa Schiebinger, *Nature's Body: Gender in the Making of Modern Science*, Beacon Press, Boston, 1993, 30.

81 William Withering, *Botanical Arrangement of British Plants*, Vol. 1, M. Swinney, Birmingham, 1788 [1776], xv, 2. The 'thread' is the stamen, and the 'shaft' is the pistil. Withering dropped the sexual terms in order to render the study of botany suitable for women, but he also wanted to translate the Latinate terms into English for a less learned audience.

82 Pulteney, *A General View of the Writings of Linnaeus*, 242–3.

83 Priscilla Wakefield, *An Introduction to Botany in a Series of Familiar Letters*, Thomas Burnfide, Dublin, 1796, 10–11. Wakefield explains that the pollen is absorbed by the pointel and vivifies the seed (12), so she did not avoid the discussion of generation altogether, but it is not described in Linnaeus's anthropomorphic terms. There was also a vehement condemnation of Linnaeus's sexual imagery, but on moral, not scientific grounds. Johann Georg Siegesbeck in Germany found the system morally repugnant and ungodly (see, for example, Larson, *Reason and Experience*, 58; Larson also notes [59] that although the early, critical reception of Linnaeus's sexual system included some condemnation of his sexual metaphors, as the system itself gained scientific acceptance criticism of the metaphors disappeared.) The English clergyman Richard Polwhele wrote a parody of Erasmus Darwin's *The Loves of The Plants* (Polwhele, *The Unsex'd Females: A Poem*, Cadell and Davis, London, 1798), in which the sexual system is said to be incompatible with female modesty (8), but it is as much, if not more, a diatribe against educated women. William Smellie Jnr. (*The Philosophy of Natural History*, Charles Elliot, Edinburgh, 1790) objected to the idea of the sexes of plants on allegedly scientific grounds, but also accused the sexualists in general of 'rashly yielding assent to the alluring seductions of analogical reasoning' (245) and accused Linnaeus in particular of having 'pushed analogy so far beyond all decent limits that it becomes truly ridiculous' (248). Many have interpreted Linnaeus's works and Erasmus Darwin's poem in the context of a much longer tradition of the association

of plants with both eroticism and sexual purity and the stereotype of the female botanist as 'sexually forward' (Sam George, *Botany, Sexuality and Women's Writing 1760–1830: From Modest Shoot to Forward Plant*, Manchester University Press, Manchester, 2007, 1). See, for example, Fleur Daugey, *Les plantes ont-elles un sexe? Histoire d'une découverte*, Éditions Ulmer, Paris, 2015; Fara, *Sex, Botany and Empire*.

84 See, for example, Mary B. Hesse, 'Models, Metaphors and Truth', in F. R. Ankersmit and J. J. A. Mooij, eds., *Knowledge and Language*, Vol. III, Kluwer, Dordrecht, 1993, 49–66.; Jacques Derrida, 'White Mythology' (1971), in Alan Bass, trans., *Margins of Philosophy*, Harvester Wheatsheaf, New York etc, 1982 [1972], 207–71.

85 Max Black, 'Metaphor', *Proceedings of the Aristotelian Society*, New Series, Vol. 55 (1954–5): 273–94; 280–1.

86 Richard Rorty, 'Hesse and Davidson on Metaphor', Richard Rorty and Mary Hesse, 'Unfamiliar Noises': I, *Proceedings of the Aristotelian Society*, *Supplementary Volumes*, Vol. 61 (1987): 283–311; 295–6.

87 Mary Hesse, 'Tropical Talk: The Myth of the Literal', Rorty and Hesse, 'Unfamiliar Noises': II, 297. To be clear, Hesse – for whom all language is metaphorical – is arguing against Rorty: 'if the claim that all language is metaphorical is true, the explanation of metaphor cannot require a basic literal language as premiss' (310).

88 Evelyn Fox Keller, *Making Sense of Life: Explaining Biological Development with Models, Metaphors and Machines*, Harvard University Press, Cambridge MA, 2002, 209. On the persistence of metaphor in scientific botany see also Schiebinger, *Nature's Body*, Chapter 1 'The Private Life of Plants'.

89 Linnaeus, *Philosophia Botanica*, §148–9, 106.

90 Thus, we cannot agree with François Delaporte's (Bachelardian) conclusion (to his *Nature's Second Kingdom*, 91) that there is in all respects a decisive break between the 'vegetality' of eighteenth-century natural history and the phytobiology or scientific plant physiology born in the nineteenth century.

Chapter 5

1 This phrase, rather than 'dibiontic life cycle', seems to be more common in botany. It originally referred to alternation of asexual and sexual generations in animals (including periodic parthenogenesis) but is now used almost exclusively in relation to plants. See David Haig, 'Homologous Versus Antithetic Alternation of Generations and the Origin of Sporophytes', *Botanical Review*, Vol. 74, No. 3 (September 2008): 395–418, 396. Roy W. Jones noted the potential for confusion in the use of the term in

1937: 'Alternation of Generations as Expressed in Plants and Animals', *Bios*, Vol. 8, No. 1 (March 1937): 19–29.

2   Sachs, *History of Botany*, 200; William Hofmeister, *On the Germination, Development, and Fructification of the Higher Crypotogamia and on the Fructification of the Coniferae*, trans. Frederick Curry, Ray Society, London, 1862 [1851], see especially 434–5.

3   Nils Svedelius, 'Alternation of Generations in Relation to Reduction Division', *Botanical Gazette*, Vol. 83, No. 4 (June 1927): 362–84, 364.

4   For readers unfamiliar with botanical terminology Michael Allaby's *Dictionary of Plant Sciences*, fourth edition, Oxford University Press, Oxford, 2019, is invaluable; for more detail Mauseth's *Botany* is an accessible general textbook even for the non-scientist. We have relied on these two books throughout.

5   Though many plants, unlike almost all animals, can have multiple and odd numbers of chromosomes in their cells ('polyploidy') without this prompting spontaneous abortion of embryos or giving rise to 'abnormalities'. Plants may also maintain different ploidies on different parts. Polyploidy arises spontaneously in wild populations but is also often the result of deliberate hybdidization in cultivars. See, for example, Hallé, *Éloge de la plante*, 280; Tassin, *À quoi pensent les plantes?*, 80; Colin Tudge, *The Secret Life of Trees*, Allen Lane, London, 2005, 22–5; Lauren Ancel Meyers and Donald A. Levin, 'On the Abundance of Polyploids in Flowering Plants', *Evolution*, Vol. 60, No. 6 (2006): 1198–206.

6   Though, as mentioned before (footnote 4), some plants do maintain alternative ploidys. We use the haploid/diploid distinction here for simplicity's sake.

7   L. H. Bailey, 'On the Untechnical Terminology of the Sex-Relation in Plants', *Science*, Friday 5 June 1896: 825–7, 826.

8   Some unfertilized eggs of some insects – for example bees – are an exception to this rule; they may develop into sterile 'workers'. See Mauseth, *Botany*, 203.

9   Ibid., 203–4.

10   For more detail see, for example, ibid., 211; Taiz et al., *Plant Physiology and Development*, 630–2.

11   Mauseth, *Botany*, 204. The microgametophyte, although 'extremely simple … is a full-fledged plant' (211); 'the megagametophyte is a distinct plant' (212). See also Taiz et al., *Plant Physiology*, 63: 'The fully cellularized embryo sac represents the mature female gametophyte or embryo sac'. P. Mascarenhas, 'The Male Gametophyte of Flowering Plants', *The Plant Cell*, Vol. 1 (July 1989): 657–64 is a good example of an essay that treats the 'morphologically simple' (663) male gametophyte as a separate plant.

12   Mauseth, *Botany*, 213. Taiz et al., *Plant Physiology and Development*, 638–9.

13 In dioecious species of flowering plants there are at least four multicellular organisms: two sporophytes and two gametophytes.

14 This account does not take into account the even more complex life cycles of some other vegetable life, for example, the red algae, which have three generations. See, for example, Mauseth, *Botany*, 475–7.

15 See, for example, Eduard Strasburger, 'The Periodic Reduction of the Number of Chromosomes in the Life-History of Living Organisms', *Annals of Botany*, Vol. 8, No. 31 (September 1894): 281–316; Sachs, *History of Botany*, 200–1; Jones, 'Alternation of Generations', 19.

16 On the use of this terminology in the early history of discussion of alternation of generations see Haig, 'Homologous versus Antithetical Alternation of Generations and the Origin of Sporophytes', 396–7.

17 L.H. Bailey, 'On the Untechnical Terminology of the Sex-Relation in Plants', *Science*, Vol. 3, No. 75 (5 June 1896): 825. Or, as John Farley puts it in 1982 (*Gametes and Spores: Ideas about Sexual Reproduction 1750–1914*, Johns Hopkins University Press, Baltimore & London, 1982, 5), the presumption of the universality of sex means the presumption of the universality of sex organs.

18 Bailey, 'On the Untechnical Terminology of the Sex-relation in Plants', 826.

19 Ibid., 827.

20 Coulter, *The Evolution of Sex in Plants*, University of Chicago Press, Chicago IL, 1914, 75.

21 Ibid., 22. Coulter's effective equation of the process of sex and the sexes is not part of the most recent definitions of 'sex'. See, for example, Leo W. Beukeboom and Nicolas Perrin, *The Evolution of Sex Determination*, Oxford University Press, Oxford, 2014, 2–14. These definitions are discussed in our Epilogue.

22 Coulter, *The Evolution of Sex in Plants*, 35.

23 Ibid., 102–3. Earlier (36–7) he says that the larger, non-motile gamete with nutritive capacity 'is recognized as the female sexual cell, or *egg*'; the other, smaller motile gamete is 'recognized as the male sexual cell, or *sperm*'.

24 Ibid., 74.

25 Ibid., 95.

26 Ibid., 93.

27 Ibid., 95.

28 Ibid., 94–5.

29 Coulter, *Plant Studies: An Elementary Botany*, D. Appleton and Company, New York, 1901, 346. He makes the same point about the early names for the megasporophyll (350). In his later *A Text-Book of Botany For Secondary Schools*, D. Appleton and Company, New York, 1910, which is a revised version of *Plant Studies*, he uses the technical vocabulary of

microsporophyll and megasporophyll but says that 'these structures also bear old names that may be used' (214). Coulter indeed drew on these old names to refer to flowers as either 'staminate' (having only stamens) or 'pistillate' (having only pistils), thus avoiding calling them 'male' or 'female'. These words are still in use today.

30 Coulter, *Plant Studies*, 348: 'the name "male cone"... should, of course, be abandoned.' See also *The Evolution of Sex in Plants*, 94. Interestingly, Ellison Hawks's *Pioneers of Plant Study* (1928), which is a popular scientific and educational text, gives a brief introductory account of plant structure, pollination and fertilization without ever referring to any part or function as 'male' or 'female' (3–15).

31 Coulter, *Plant Studies*, 348. Scientist-turned-historian-of-science John Farley makes much the same point in his *Gametes and Spores* (100): 'Even modern day botanists have failed to come to terms with Hofmeister's work. By retaining an essentially Linnean terminology, they have managed to confuse generations of botany students. The male sex organs of lower plants are still called "antheridia", meaning "anther-like"; but this is a misnomer.' Similarly, 'Cones are referred to as male and female, but in reality they produce only asexual spores. An even more confusing distinction is the labelling of stamens and anthers as "male-producing organs" and pistils as "female-producing organs". As became apparent over a century ago, however, both stamens and pistils produce spores; they are not sexual organs.'

32 Bailey, 'On the Untechnical Terminology of the Sex-Relation in Plants', 826–7.

33 Charles J. Chamberlain, 'Alternation of Generations in Animals from a Botanical Standpoint', *Science*, Vol. 22, No. 555 (18 August 1905): 137–44, 142. In the textbooks that Coulter and Chamberlain co-authored Coulter's position won out, perhaps because he was senior to Chamberlain in the botany department at the University of Chicago. See John M. Coulter and Charles J. Chamberlain, *Morphology of Gymnosperms*, University of Chicago Press, Chicago, 1910 [1901]; John M. Coulter and Charles J. Chamberlain, *Morphology of Angiosperms*, D. Appleton and Company, New York, 1903. More recently Alexander P. Dyachenko ('Botanical Terminology: New Twists or Tradition?', *Skvortsovia*, Vol. 3, No. 3 (2017): 122–9) has argued much the same as Chamberlain on the basis that these, like many other botanical terms, are 'conditional' and should be retained even if they 'do not fit with reality' (123).

34 See, for example, Beukeboom and Perrin, *The Evolution of Sex Determination*, 10: 'Male and female function refers to production of small and large gametes respectively (sperm and ova). At this haploid and unicellular level, there are two sexes, and only two.' Beukeboom and Perrin also refer to the multicellular individuals 'allocated' to the male and female functions as 'sexes' (but also, confusingly, 'genders') (10). However, the distinctions are not carried through strictly in relation to plants: 'Individuals gametophytes or sporophytes specializing strictly in one or the other sexual function will be referred to as male and females respectively' (i.e. the strict botanical distinction between sporophyte and gametophyte is not in fact respected). Other textbooks give circular definitions, for example Eldon D. Enger, Frederick C. Ross and David B. Bailey, *Concepts in Biology*, twelfth edition, McGraw Hill, Boston etc, 2007, 596: 'Sex is the nature of the biological difference between males and females.'

35 Mauseth, *Botany*, 206, 218 (though he glosses dioecy as 'similar to the condition of separate gender in mammals'), 548; and Chapters 9, 16, and 19–23 generally. Peter R. Bell ('The Alternation of Generations', in J.A. Callow, ed., *Advances in Botanical Research*, Vol. 16, Academic Press, London, 1989, 55–93) speaks of asexual and sexual generations, of 'the spores giving rise to the sexes' and of the sex of gametophytes, quietly avoiding sexing the sporophyte.

36 Mauseth, *Botany*, 206, 209. There is one reference to the ancestral 'bisexual' flower (549). In the chapter on 'Angiosperm Reproduction and Biotechnology' in Neil A. Campbell et al's general biology textbook (*Biology*, eighth edition, Pearson Benjamin Cummings, San Francisco, CA, 2008, 801–20) 'male' and 'female' are only used of gametophytes. Flowers are referred to as 'staminate' and 'carpellate'.

37 See, for example, Heslop-Harrison, 'The Angiosperm Stigma', 59–68 and H. Kaufmann, H. Kirsch, T. Wemmer, A. Peil, F. Lottspeich, H. Ulrig, F Salamini and R. Thompson, 'Sporophytic and Gametophytic Self-Incompatibility', in M. Cresti and A. Tiezzi, eds., *Sexual Plant Reproduction*, Springer-Verlag, Berlin etc., 1992, 115–25. Neither of these essays refers to 'male' and 'female' at all. Jannathan Mamut, Ying-Ze Xiong, Dun-Yan Tan and Shuang-Quan Huang ('Pistillate Flowers Experience More Pollen Limitation and Less Geitonogamy Than Perfect Flowers in a Gynomonoecious Herb', *The New Phytologist*, Vol. 201, No. 2 (January 2014): 670–7) clearly prefer to avoid 'male' and 'female' throughout, but gloss 'pistillate', 'staminate' and 'perfect' as 'female', 'male' and 'hermaphrodite' respectively (670).

38 See, for example, Sophie C. Ducker and R. Bruce Knox, 'Pollen and Pollination: A Historical Review', *Taxon*, Vol. 34, No. 3 (August 1985): 401–19.

39 For example, a recent book from Kew Botanical Gardens, *Flora: Inside the Secret World of Plants,* Dorling Kindersley, London, 2018, aimed at an adult audience, labels the 'male' and 'female' parts of the flower in almost the same way as Richard and Louise Spilsbury's, *The Life of Plants: Plant Reproduction,* Heinemann, Portsmouth NH, 2008, which is for reading age eight to eleven. It is also not surprising to find some bowdlerization in the popular literature that trades (at least in part) on the invocation of human sexuality for its appeal. See, for example, Bastiaan Meeuses and Sean

Morris, *The Sex Life of Flowers*, Faber and Faber, London/Boston, 1984; Daugey, *Les plantes ont-elles un sexe?*, 20, 28.

40 See, for example, Mancuso, *Brilliant Green*, 105–6, where stamens and pistils are described as male and female reproductive organs and the pollen granule is said to be 'the male gamete (male seed)'.

41 This has been the case for some time now. See, for example, F. O. Bower, *Botany of the Living Plant*, fourth edition, Macmillan & Co, London, 1947 [1919], 257; R. Frankel and E. Galun, *Pollination Mechanisms, Reproduction and Plant Breeding*, Springer-Verlag, Berlin, 1977, 10; Andrew G. Stephenson and Robert I. Bertin, 'Male Competition, Female Choice, and Sexual Selection in Plants', in Leslie Real, ed., *Pollination Biology*, Academic Press, Orlando etc, 1983, 109–49, 112; Taiz et al., *Plant Physiology and Development*, 625 ff.

42 Attenborough, *The Private Life of Plants*, 95. See also Michael Proctor, Peter Yeo and Andrew Lack, *The Natural History of Pollination*, HarperCollins, London, 1996, 27: staminate and pistillate flowers are 'often loosely referred to as "male" and "female" flowers'.

43 This is not to say that plant scientists are wrong, exactly, when they do this. Nevertheless, the ubiquity of this terminological practice in the scientific and popular scientific literature is interesting and the implications of this dual discourse need to be considered. What is the significance of the fact that the use of strictly speaking 'incorrect' terminology is not considered to be wrong?

44 Spencer C. H. Barrett and Deborah Charlesworth, 'David Graham Lloyd, 20 June 1937–30 May 2006', *Biographical Memoirs of Fellows of the Royal Society*, Vol. 53 (2007): 203–21, 207. On Lloyd's career and contribution to the plant sciences see also Colin J. Webb, 'David Graham Lloyd', *New Zealand Journal of Botany*, Vol. 45 (2007): 443–9; and Spencer C. H. Barrett and Laurence D. Harder, 'David G. Lloyd and the Evolution of Floral Biology: From Natural History to Strategic Analysis', in Harder and Barrett, eds., *Ecology and Evolution of Flowers*, Oxford University Press, Oxford, 2006, 1–21.

45 P. R. Neal and G. J. Anderson discuss the confusion that still surrounds this terminology in 'Are "Mating Systems" "Breeding Systems" of Inconsistent and Confusing Terminology in Plant Reproductive Biology? Or is it the Other Way Around?', *Plant Systems and Evolution*, Vol. 250 (2005): 173–85.

46 Darwin, *The Different Forms of Flowers on Plants of the Same Species*, D. Appleton, New York, second edition, 1899 [1877], 11–12. He also identifies two subgroups of polygamous plants, according to whether the three different forms of flowers are found in the same plant (monoecious polygamy) or on different plants, the latter called 'trioicious'.

47 Ibid., 11.

48 Ibid., 18–21.

49 Ibid., 24–8. At the beginning of the discussion Darwin writes that 'The fertlisation of either form with pollen from the other form may be conveniently called a *legitimate union*, from reasons hereafter to be made clear; and that of either form with its own-form pollen an *illegitimate union*.' (24)

50 Ibid., 28.

51 More recent research suggests that heterostyled (or 'distylous') species of *Linum* are in fact wholly 'self-incompatible', that is, unable to be fertilized after self- or intra-morph pollination. See Brian G. Murray, 'Floral Biology and Self-Incompatibility in Linum', *Botanical Gazette*, Vol. 147, No. 3 (September 1986): 327–33. Darwin was perfectly well aware that the conditions under which he conducted his experiments meant that he could never be sure exactly which plant had pollinated which. But the argument that we are constructing here does not rest or fall on Darwin's results; the conceptual issues remain the same.

52 Darwin, *The Different Forms*, 90.

53 Ibid., 244, 267, 275, 277.

54 Ibid., 137–8.

55 Ibid., 159; for the discussion of the experiments see 143–61.

56 Ibid., 149, 138.

57 Ibid., 163.

58 Ibid., 164.

59 Whereas, he says, hermaphrodite animal individuals can fertilize or be fertilized by any other individual of their species, 'the essential character of plants belonging to the heterostyled class is that the individuals are divided into two or three bodies, like the males and female of dioecious plants or of the higher animals, which exist in approximately equal number and are adapted for reciprocal fertilisation'. Darwin, *The Different Forms*, 246. What Darwin is identifying are in fact 'mating types' not strictly speaking sexes, to the extent that the latter are by definition only two. (We discuss the difference between mating types and sexes in the Epilogue.) One of the most important conclusions, for Darwin, is that these plants show that the definition of species in terms of interfertility is shown by these plants to be inadequate, because members of the same species cannot necessarily produce fertile offspring even if they can 'mate'. Darwin, *The Different Forms*, 247.

60 Lloyd, 'Breeding Systems in Cotula L. (Compositae, Anthemideae). I. The Array of Monoclinous and Diclinous Systems', *The New Phytologist*, Vol. 71, No. 6 (November 1972): 1181–94, 1181. 'Monoclinous' flowers have stamens and carpels in the same flower; 'diclinous' plants have stamens and carpels in different flowers on the same plant.

61 Is it a coincidence that this terminological shift is contemporary with the feminist popularization of the sex/gender distinction in the Anglophone psychological and social sciences and the humanities? It would be foolish to hazard a claim about Lloyd's intentions in this regard, but, in both cases the shift to 'gender' gives notice of a conceptual innovation, not simply a terminological alternative. Renee Borges ('Gender in Plants', *Resonance*, Vol. 3, No. 11 (November 1988)) discusses 'the difference between plant sex and plant gender' with reference to Lloyd's work, but proposes a different meaning for 'sex' (for her this means 'mating systems') than in this chapter.

62 Richard B. Primack and David G. Lloyd, 'Sexual Strategies in Plants IV. The Distributions of Gender in Two Monomorphic Shrub Populations', *New Zealand Journal of Botany*, Vol. 18, No. 1 (1980): 109–14, 109.

63 Lloyd, 'Breeding Systems in Cotula L. I', 1182. Lloyd continues: 'The customary definitions of diclinous conditions ignore populations which do not fit easily into one of the morphologically defined groups, and are inadequate for a detailed study of breeding systems.'

64 Ibid., 1183.

65 David G. Lloyd, 'Sexual Strategies in Plants III: A Quantitative Method for Describing the Gender of Plants', *New Zealand Journal of Botany*, Vol. 18, No. 1 (1980): 103–8, 103.

66 Ibid., 105.

67 Lloyd gives formulas for the estimation of functional gender, which basically employ estimates of maternal and paternal expectations with various assumptions, for example that all androecial units have the same probability of contributing genes to the next generation. See Lloyd, 'Sexual Strategies in Plants III', 105–6.

68 But even here there may be differences. For example, a flower with a functional gynoecia and non-functional androecia would, on the traditional method, be seen as an anomaly, and called a 'sterile hermaphrodite'. According to the functional concept of gender, however, it is straightforwardly a female because, as Lloyd says ('Breeding Systems in Cotula L… I', 1183), it is neither sterile not an hermaphrodite.

69 David G. Lloyd, 'Theoretical Sex Ratios of Dioecious and Gynodioecious Angiosperms', *Heredity*, Vol. 32, No. 1 (1974): 11–34, 29.

70 See, for example, Lloyd, 'Breeding Systems in Cotula L… I', 1182–3; Lloyd, 'Theoretical Sex Ratios of Dioecious and Gynodioecious Angiosperms', 13 ('the males of gynodioecious populations contribute a minor but appreciable proportion of their genes via ovules'), 29; David G. Lloyd, 'Female-Predominant Sex Ratios in Angiosperms', *Heredity*, Vol. 32, No. 1 (1974): 35–44, 36 ('In gynodioecious populations males produce some offspring from seed'); David G. Lloyd, 'The Transmission of Genes via Pollen and

Ovules in Gynodioecious Angiosperms', *Theoretical Population Biology*, Vol. 9, No. 3 (1976): 299–316, 300, 311, 312; David G. Lloyd, 'Parental Strategies of Angiosperms', *New Zealand Journal of Botany*, Vol. 17 (1979): 595–606, 596; David G. Lloyd and K. S. Bawa, 'Modification of the Gender of Seed Plants in Varying Conditions', *Evolutionary Biology*, Vol. 17 (1984): 255–338, 259, 261; Lynda F. Delph and David G. Lloyd, 'Environmental and Genetic Control of Gender in the Dimorphic Shrub Hebe subalpina', *Evolution*, Vol. 45, No. 8 (December 1991 (1957–64), 1957).

71 Lloyd, 'The Transmission of Genes via Pollen and Ovules in Gynodioecious Angiosperms', 300. Lloyd further specifies that '[t]he extremes of this range, strictly unisexual plants, are referred to as constant males and female, while individual plants (or sexes generally) contributing only predominantly by gametes of one kind are described as inconstant…. In gynodioecious populations, some or all males are appreciably inconstant while female are virtually or strictly constant'. See also Lloyd, 'The Distribution of Gender in Four Angiosperm Species Illustrating Two Evolutionary Pathways to Dioecy', *Evolution*, Vol. 34, No. 1 (January 1980): 123–34, 123: 'In angiosperms, gender is a quantitative phenomenon which can be measured on a continuous scale between the strictly male and the strictly female extremes.'

72 Primack and Lloyd, 'Sexual Strategies in Plants IV', 113. See also Lloyd, 'The Distribution of Gender in Four Angiosperm Species', 130.

73 Lloyd, 'Sexual Strategies in Plants III', 107.

74 Primack and Lloyd, 'Sexual Strategies in Plants IV', 109. See also Lloyd, 'Sexual Strategies in Plants III', 104: 'The discrete, typological nature of the morphological terms does not lend itself readily to the recognition of intermediate conditions or to the description of differences in the relative pollen and ovule contributions of flowers and individuals of the same morphological type. Morphological descriptions… ignore the fact that the sexual performance of a flower of plant depends not only on its own nature, but also on the gametes produced by other flowers and plants in the same population.'

75 'Cosexual' is Lloyd's term for a monomorphic population, that is, where there is only one type of plant, with both male and female function, which on average contributes equally through both functions. It is distinguished from hermaphroditism in referring specifically to function, not morphology. See, for example, Lloyd, 'Parental Strategies of Angiosperms', 596; Lloyd, 'Sexual Strategies in Plants III', 107; Lloyd, 'The Distribution of Gender in Four Angiosperm Species', 130.

76 Lloyd, 'Sexual Strategies in Plants I', 74.

77 Lloyd, 'Breeding Systems in Cotula L…. I', 1183: 'The distinction between the classes of breeding systems are not always absolute. It is necessary to decide, for example, whether occasional male-sterile flowers (or plants) warrant the recognition of gynomonoecy (or of sexual dimorphism). In both cases the flowers or plants which occur less frequently are recognized as a significant element of the breeding system only if they are sufficiently frequent or persistent to meaningfully affect the pattern of mating. Decisions made on this basis may sometimes be arbitrary, even after close examination, but they are at least based on consideration of the biological significance of the variation and not on rigid typological criteria.' Lloyd notes that 'there are frequent and diverse intergrades between the major sex conditions, as Darwin (1877 [*The Different Forms of Flowers on Plants of the Same Species*]) pointed out' ('Sexual Strategies in Plants III', 103).

78 Lloyd, 'Parental Strategies of Angiosperms', 597.

79 Primack and Lloyd, 'Sexual Strategies in Plants IV', 109. In 'Sexual Strategies in Plants III' (105, 107) Lloyd contrasts the 'effective gender' of a plant ('the relative success that plant eventually has in leaving descendants through its male and female gametes'), which may not in fact ever be actually known, with the relative investments of a plant on gynoecia and androecia, which are comparatively easy to record.

80 See Barrett and Harder, 'David G. Lloyd', 10: 'The hallmark of Lloyd's work on plant sexual systems was his elaboration of the concept of gender. In contrast to the term "sex," which reflects phenotype, "gender" describes the relative genetic contribution of individuals to the next generation as female and male parents, of their functional "femaleness" or "maleness".' In the introductory essay 'Gender and Sexual Dimorphism in Flowering Plants: A Review of Terminology, Biogeographic Patterns, Ecological Correlates and Phylogenetic Approaches', in M. A. Geber, T. E. Dawson and L. F. Delph, eds., *Gender and Sexual Dimorphism in Flowering Plants*, Springer, Berlin, 1999, Ann K. Sakai and Stephen G. Weller also specify this use and understanding of the distinction: throughout the book, they say, 'we use *gender* to describe the functional sex expression of the plant (or flower) and *sexual dimorphism* to refer to differences (often morphological) between two classes of plants in primary or secondary characters (1). However, they use the more traditional term 'hermaphrodite' for plants with both male and female function in monomorphic or dimorphic populations to avoid what they call the 'counter intuitive' (8) idea that males can produce significant amounts of seeds. But the 'intuition' to which they want to stay true belongs to the animal model.

81 Lloyd, 'Sexual Strategies in Plants III', 103.

82 See ibid., 107.

83 Lloyd, 'The Transmission of Genes via Pollen and Ovules in Gynodioecious Angiosperms', 312.

84 Lloyd, 'The Distribution of Gender in Four Angiosperm Species', 130; Lloyd, 'Sexual Strategies in Plants III', 104.

Chapter 6

1 See Farley, *Gametes and Spores*, 9–14.

2 On the use of the term 'indigenous' see our Introduction, note 11.

3 Hans Werner Ingensiep, 'The History of the Plant Embryo. Terminology and Visualization from Ancient until Modern Times', *Hist. Phil. Life Sci.*, Vol. 26 (2004): 309–31; 311.

4 Aristotle, *Generation of Animals*, I 23, 731a1–14, 1134.

5 Cesalpino, *De plantis libri XVI*, Chapter I, 8; see also VIII, 16.

6 Ingensiep, 'The History of the Plant Embryo', 312.

7 Grew, *The Anatomy of Plants*, Book IV, Chapter V, 206. He refers to the 'uterus' in the (unpaginated) Explication of Tables (Table LVII).

8 Malpighi, Marcelli, *Anatome Plantarum*, Regiae Societati, Londini, 1675, 57. See Ingensiep, 'The History of the Plant Embryo', 312–14. Morton (*History of Botanical Science*, 185) notes the acuity of many of Malpighi's observations but because of his 'elaborate analogy', Malpighi 'confused himself by attempting to see structures present in the animal uterus and not all his figures are free from errors of observation due to this cause'.

9 See Ingensiep, 'The History of the Plant Embryo', 311, 315. To take just two examples, Abbé Lazarro Spallanzani, *Dissertations Relative to the Natural History of Animals and Vegetables*, J. Murray, London, 1789 [English translation of *Experiencias Para Servir a La Historia de La Generación De Animales y Plantas*, 1786], unpaginated index of contents: 'the embryo does not at all depend for its existence upon the powder of the stamina; therefore, secondly, the embryo exists in the ovarium independently of this powder; thirdly, nor is it the result of two principles, one depending on the pollen, and the other upon the pistil, as others suppose'. Erasmus Darwin, on the other hand, writes (*Zoonomia, Or, The Laws of Organic Life*, J. Johnson, London, 1794, 485): 'That the embryon is secreted or produced by the male and not by the conjunction of fluids from both male and female, appears from the analogy of vegetable seeds.'

10 See Sachs, *History of Botany*, 124–5: 'Gärtner's theory of the seed is one of his most important contributions to the science…. His giving the term embryo to that part of the seed which is capable of development was also an advance in respect of logic and morphology … What Gärtner now named embryo, had been up to his time called the "corculum seminis"'. On the idea of the 'corculum seminis' see Ingensiep, 'The History of the Plant Embryo', 316–17.

11 See Donald Alexander Johansen, *Plant Embryology: Embryogeny of the Spermatophyta*, Chronica Botanica Company, Waltham, MA, 1950, 5: in 1827 Adolphe-Théodore Brongniart 'recognised and named the "sac embryonnaire," substituting this term for the "amniotic sac" of Malphigi'.

12 Hooper claims, for example, 'That the vessels of the bud which inosculate with the trunk of the tree, are to the bud what those of the placenta and cotyledons are to animals'. Hooper, *Observations on the Structure and economy of Plants*, 117. See also 47, 101, 115, 116. For just one earlier example of an obvious analogy between plant part and animal placenta, see Henry Baker, 'The Discovery of a Perfect Plant in Semine', *Philosophical Transactions*, Vol 41 (1739–41): 448–55, 451.

13 The plant placenta is mentioned in Robert Brown, 'Observations on the Structure and Affinities of the More Remarkable Plants Collected by... Walter Oudney etc', London, 1826, reprinted in Brown, *Miscellaneous Botanical Works*, Vol. I, Ray Society, London, 1866; see 270–2, for example.

14 John Lindley Thomas Moore, *The Treasury of Botany: A Popular Dictionary of the Vegetable Kingdom; with Which Is Incorporated a Glossary of Botanical Terms*, Longmans, Green and Co., London, 1866, 899. See also, for example, Sachs, *Lehrbuch der Botanik*, 482; Allaby, *A Dictionary of Plant Sciences*, 412: 'Placenta. In flowers, the part of the ovary wall formed from the fused margins of the carpel or carpels, on which are carried the ovules'.

15 Which is why Ferdinand Von Mueller objected to the use of the term 'ovary' in botany: 'because we never employ the term ovum for the first stage of the seed of any plant. The inconsonance of the combination of ovary with ovule has long been perceived'. Ferdinand Von Mueller, 'Considerations of Phytographic Expressions and Arrangements', *Journal and Proceedings of the Royal Society of New South Wales*, Vol. 22 (1888): 187–204, 193. It is not clear that Mueller's proposed alternative, 'ovulary' (194), quite clears the problem up.

16 These might be translated as 'seed', 'seed sac', 'fruit node' and 'seed enclosure'.

17 Hallé, *Éloge de la plante*, 188–9.

18 Michèle Le Doeuff, *The Philosophical Imaginary*, trans. Colin Gordon, Continuum, London and New York, 2002 [1980], 7.

19 Ibid., 5, 6.

20 See, for example, Taiz et all, *Plant Physiology and Development*, 626–31.

21 Which immediately produces conceptual problems: Does the mother survive the division? Trewavas (*Plant Behaviour and Intelligence*) refers to the pluripotent cells of root and shoot meristems as undifferentiated embryonic areas (6, 47) and to embryonic tissue in the cambium (7), but this is unusual.

22 Meeuse and Morris, *The Sex Life of Flowers*, 15, 8.

23 Hallé, *Éloge de la plante*, 187. Hallé stresses the parasitic relation of the gametophyte to the sporophyte, which, he notes, gives rise to 'this apparent absurdity: a plant that parasitizes itself!' (187).

24 E. J. H. Corner, *The Life of Plants*, University of Chicago Press, Chicago and London, 2002 [1964], 115.

25 As in the title of David Haig and Mark Westoby, 'Inclusive Fitness, Seed Resources, and Maternal Care', in Jon Lovett Doust and Lesley Lovett Doust, eds., *Plant Reproductive Ecology: Patterns and Strategies*, Oxford University Press, Oxford, 1988, where the main issue is 'maternal investment' (a plant's investment of energy in carpels and seed production) and associated behaviours, for example, 'seed abortion'.

26 See, for example, David G. Lloyd, 'Sexual Strategies in Plants I: An Hypothesis of Serial Adjustment of Maternal Investment during One Reproductive Season', *The New Phytologist*, Vol. 86 (1980): 69–79. Tassin (*À quoi pensent les plantes?*, 15) compares the totality of the plant biosphere to 'our second womb [*matrice*]'.

27 Suzanne W. Simard, David A. Perry, Melanie D. Jones, David D. Myrold, Daniel M. Durall and Randy Molina, 'Net Transfer of Carbon between Ectomycorrhizal Tree Species in the Field', *Nature*, Vol. 388 (7 August 1997): 579–82. Simard's experiments 'labelled' the seedlings with radioactive carbon isotopes so that the movement of the products of photosynthesis (carbohydrates, or sugars) between plants could be tracked. Obviously, the seedlings in shade could photosynthesize less and thus produce less carbohydrate, making them the 'sinks' in these experiments. The experiments are described in Simard's memoir, *Finding the Mother Tree: Uncovering the Wisdom and Intelligence of the Forest*, Allen Lane, London, 2021, 142–61.

28 See Simard, *Finding the Mother Tree*, 175–6.

29 See, for example, Kevin J. Beiler, Daniel M. Durall, Suzanne W. Simard, Sheri A. Mazwell and Annette M. Kretzer, 'Architecture of the Wood-Wide Web: *Rhizopogon* spp. Genets Link Multiple Douglas-fir Cohorts', *New Phytologist*, Vol. 185 (2010): 543–53; Simard, Kevin J. Beiler, Marcus A Bingham, Julie R. Deslippe, Leanne J. Philip and François P. Teste, 'Mycorrhizal Networks: Mechanisms, Ecology and Modelling', *Fungal Biology Reviews*, Vol. 26 (2012): 39–60; Monika A. Gorzelak, Amanda K. Asay, Brian J. Pickles and Suzanne Simard, 'Inter-Plant Communication through Mycorrhizal Networks Mediates Complex Adaptive Behaviour in Plant Communities', *AoB Plants*, Vol. 7 (2015): 1–13.

30 See, for example, Suzanne W. Simard, 'Mycorrhizal Networks and Complex Systems: Contributions of Soil Ecology Science to Managing Climate Change Effects in Forested Ecosystems', *Can. J. Soil Sci.*, Vol. 89 (2009): 369–82. Simard writes (*Finding the Mother Tree*, 214) that moving from the forestry service to a university job allowed her 'to get at the basic questions of relationships in the forest, which had deepened from ideas about connection and communication between trees to a more holistic understanding of forest intelligence'.

31 See, for example, Suzanne W. Simard, 'Mycorrhizal Networks Facilitate Tree Communication, Learning and Memory', in F. Baluska, M. Gagliano and G. Witzany, eds., *Memory and Learning in Plants*, Springer, Cham, Switzerland, 2018: 191–213.

32 See, for example, Marcus A. Bingham and Suzanne W. Simard, 'Ectomycorrhizal networks of *Pseudotsuga menziesii* var. *glauca* [Rocky Mountain Douglas Fir] Trees Facilitate Establishment of Conspecific Seedlings under Drought', *Ecosystems*, Vol. 15 (2012): 188–99.

33 Simard et al., 'Architecture of the Wood-Wide Web', 549, 550.

34 https://mothertreeproject.org/. A full list of Simard's publications and other relevant research can be found on the Mother Tree Project website.

35 Simard, *Finding the Mother Tree*, 47–8.

36 Ibid., 75, 77.

37 Ibid., 189, 190.

38 Ibid., 242. In a short film (*Mother Trees Connect the Forest*, directed by Dan McKinney and Julia Dordel, 2011), Simard says of a massive, 500-year-old tree: 'You can think of this as a Mother Tree in the sense that it is a dominant tree in the forest, it's probably networked into all of the trees all around it, although they are of different species.' For the purposes of various experiments seedlings in lab experiments are also designated 'Mother Trees' (*Finding the Mother Tree*, 268, 286–7) when the point is to discover what they give or communicate to other seedlings planted nearby.

39 Simard, *Finding the Mother Tree*, 228. Over the page, having likened the forest network to a neural network, Simard continues (230): 'The fungal hyphae, too, perceive their environment and alter their architecture and physiology. Like parents and children, my girls and Don [Simard's husband] and me, adapting to change, aligning to learn new things, figuring out how to endure. I'd be home tonight. *Mothering…* I felt a kinship with the Mother Trees, grateful for accepting me and giving me these insights.' See also 287: 'I imagined the flow of energy from the Mother Tree as powerful as the ocean tide, as strong as the sun's rays, as irrepressible as the wind in the mountains, as unstoppable as a mother protecting her child. I knew that power in myself even before I'd uncovered these forest conversations.'

40 Ibid., 277. Retrospectively Simard's research questions are even presented in anthropomorphic terms: 'Why do [trees] have human-like behaviours, and why do they work like civil societies?' (Ibid., 5).

41 Wohlleben, *The Hidden Life of Trees*, 33, 34.

42 Wohlleben in the film *Intelligent Trees*, directed by Julia Dordel and Guido Tölke, 2016.

43 In Dordel and Tölke's *Intelligent Trees* Simard notes that 'it is a matter of language… I can use that language'. In *Finding the Mother Tree* (274), speaking of a TEDYouth talk, Simard writes: 'I struggled with anthropomorphisms that I knew would be criticized by scientists, but I chose to use terms such as "mother" and "her" and "children" anyway to help the kids understand the concepts.' Richard Powers's novel *The Overstory* (Vintage, London, 2018) presents the reader with a fictionalized plant scientist and science popularizer in the figure of Patricia

(Patty) Westerford, based on Simard. Patty discovers the underground communicative networks of trees but is ridiculed and professionally ostracized for her ideas. After some years off grid, Patty's ideas are rediscovered by a new generation of researchers, and Patty, now fêted, takes up her scientific research again and writes a bestselling popular scientific book. Remarking on her use of phrases such as 'giving trees' the narrative imagines Patty's response: 'The reading public needs such a phrase to make the miracle a little more vivid, visible. It's something she learned long ago from her father: people see better what looks like them. *Giving trees* is something any generous person can understand and love' (276).

44 Which is why the capitalist market metaphors, so prevalent in the plant advocacy literature, seem so curiously ill-judged. To cite just one example, Mancuso (*Brilliant Green*, 58) compares a plant root growth into mineral rich soil to a mining company investigating substantial resources to open new galleries, counting on the future revenues it anticipates earning'. Plant pollination is 'a huge market' in which insects pay for commodities (nectar) with their labour; although nectar is also said to be the plants' unique currency (110).

45 Simard, *Finding the Mother Tree*, 294. See also 66, 76, 281–3, 293–4. Simard refers to ideas common amongst the indigenous peoples of what is now British Columbia, including the Coast Salish, Secwepemc and Heiltsuk peoples.

46 Wohlleben, in *Intelligent Trees*, dir. Dordel and Tölke.

47 http://www.unesco.org/new/en/natural-sciences/priority-areas/links/related-information/what-is-local-and-indigenous-knowledge

48 Robin Wall Kimmerer, *Braiding Sweetgrass: Indigenous Wisdom, Scientific Knowledge and the Teachings of Plants*, Penguin, London, 2013, 44, 37.

49 Ibid., 346. Wall Kimmerer is the director of the Center for Native Peoples and the Environment (SUNY College of Environmental Science and Forestry).

50 Ibid., 8. Here 'us' refers both to people of indigenous descent, like Wall Kimmerer, whose ancestors had been forcibly removed from their lands and alienated from their traditional cultures, and to all of us – or those of us who would like to understand how to learn with indigenous knowledge.

51 Ibid., 49–55.

52 Ibid., 56, 249.

53 Ibid., 249.

54 Wall Kimmerer, *Braiding Sweetgrass*, 250. In some of the small and often private ceremonies that she describes performing as she, for example, gathers plants for food or for the purposes of research, Wall Kimmerer exemplifies this kind of creation and practice.

55 This advantage is also increasingly apparent in various wilding or rewilding projects. See, for example, Isabella Tree, *Wilding: The Return of Nature to a British Farm*, Picador, London, 2018.

56 Dale Turner, *This Is Not a Peace Pipe: Towards a Critical Indigenous Philosophy*, University of Toronto Press, Toronto, 2006, 99.

57 Turner, *This Is Not a Peace Pipe*, 100.

58 Ibid., 116, 115.

59 Ibid., 99.

60 Philippe Descola, *Beyond Nature and Culture*, trans. Janet Lloyd, University of Chicago Press, Chicago, 2013 [2005].

61 Eduardo Viveiros de Castro, *Cannibal Metaphysics: For a Post-Structural Anthropology*, trans. Peter Skafish, University of Minnesota Press, Minneapolis/London, 2017 [2009], 40–1. Thus, for Viveiros de Castro, Narcissus is the 'patron saint or tutelary spirit of anthropology'; Part One of *Cannibal Metaphysics* is called 'Anti-Narcissus'.

62 Ibid., 42.

63 Ibid., 187, 188, 190.

64 'The philosophy of Deleuze… is where I found the most appropriate machine for retransmitting the sonar frequency that I had picked up from Amerindian thought.' Ibid., 92.

65 Ibid., 188, 191.

66 Ibid., 194.

67 Ibid., 188, 196. This seems to offer a way to address Turner's worry about whether indigenous philosophy can be articulated in English (*This is Not a Peace Pipe*, 116). Perhaps 'indigenous philosophy proper', as Turner puts it, can be translated into Western concepts *so long as those concepts are able to be transformed.*

68 Viveiros de Castro, 'Cosmological Deixis and Amerindian Perspectivism', *The Journal of the Royal Anthropological Institute*, Vol. 4, No. 3 (September 1998): 469–88, 470.

69 Viveiros de Castro, *Cannibal Metaphysics*, 56–7.

70 Ibid., 68, 69. Eduardo Kohn (*How Forest Think: Toward an Anthropology Beyond the Human*, University of California Press, Berkeley, 2013) identifies a similar kind of perspectivism amongst the Ávila Runa people of Upper Amazonian Ecuador. Kohn explains that in Quichua 'Runa' means '(human) person'; 'puma' means 'jaguar' or 'predator'. The people in Ávila speak to Kohn of the 'runa puma' 'shape-shifting human-jaguars… beings who can see themselves being seen by jaguars as fellow predators, and who also sometimes see other humans the way jaguars do, namely, as prey' (2). Kohn presents Ávila Runa perspectivism as a 'stance' that 'assumes a fundamental similarity among selves' (95), be those selves, humans, dogs, jaguars, vultures or (tantalizingly; Kohn does not develop this beyond a mere mention) plants (75).

71 Castro, *Cannibal Metaphysics*, 70.

72 Ibid., 72.

73 Ibid., 66, 69. As Viveiros de Castro also observes: 'If nothing prevents an existent from being conceived of as a person – as an aspect, that is, of a biosocial multiplicity – nothing else prevents another human collective from *not* being considered one. This is, moreover, the rule. The strange generosity that makes peoples like Amazonians see humans concealed under the most improbable forms or, rather, affirm that even the most unlikely of beings are capable of seeing themselves as humans is the double of the well-known ethnocentrism that leads these same groups to deny humanity to their fellow men [*congénères*] and even (or above all) to their closest geographical or historical cousins' (58).

74 Ibid., 69. Viveiros de Castro calls this a 'controlled equivocation' (87).

75 Ursula K. Le Guin, 'Vaster Than Empires and More Slow' [1971], in *The Wind's Twelve Quarters*, Victor Gollancz, London, 1989, 196.

76 Ibid., 63. '"Human" is a term designating a relation, not a substance.' 'Human' is more a pronoun, less a substantive (59). Kohn makes the same point in *How Forests Think* (200): '"Runa" ["(human) person"]... marks a relational subject position in a cosmic ecology of selves in which all beings see themselves as persons. "Runa" here is the self, in continuity of form. All beings are, from their points of view, in a sense "Runa", because this is how they would experience themselves when "saying" I.'

77 Viveiros de Castro, *Cannibal Metaphysics*, 192.

78 Wall Kimmerer, *Braiding Sweetgrass*, 19, 57. Hall, *Plants as Persons*, also contrasts indigenous animisms from Australia, North America and Aotearoa/New Zealand with the Western, anthropocentric view of plants. See his Chapter Five, especially.

79 Tassin, *Penser comme un arbre*, 59, 58, 126. Hallé (*Éloge de la plante*, 15) writes that 'we are attracted by what resembles us, but remain happily indifferent to what does not reflect our image back to us'. But as this is the *effect* of zoocentrism and anthropomorphism it will not be corrected by more and more of the same.

80 To be fair, Tassin's invocation of alterity does not stop him from developing a rich account of plant (especially tree) existence.

81 Viveiros de Castro, *Cannibal Metaphysics*, 85–6.

82 Ibid., 87.

83 Admittedly this is more an aspiration than an achievement in *Cannibal Metaphysics*, but it would be naïve to expect otherwise. Peter Skafish's introduction to the English edition of *Cannibal Metaphysics* (9–33) is helpful in bringing this aspect to the fore. Kohn's ambition is similarly large but promissory. He claims that his anthropology beyond the human 'changes our understanding of foundational analytical concepts such as context but also others, such as representation, relation, self, ends, difference, similarity, life, the real, mind, person, thought, form, finitude, future, history, cause, agency, relation, hierarchy, and generality' (*How*

Forests Think, 22–3). Perhaps it would have been better to say that this is what it aims to do.

84 Viveiros de Castro, *Cannibal Metaphysics*, 87, 89.

85 Ibid., 90. He continues: 'comparing the commensurate, I think, is a task best left to accountants'.

86 See Hallé, *Éloge de la plante*, Chapter 5, 'L'évolution'. 181–264. On his comparative method see ibid., 13, 22–3, 36–7.

87 Burgat, *Qu'est-ce qu'une plante?*, 18, 53, 89. Burgat admires Tassin's resolutely anti-anthropomorphic approach but describes his explanations of plant phenomena as reductively physicalist (55, 90–1).

88 Burgat, *Qu'est-ce qu'une plante?*, 18, 75, 87.

89 Ibid., 91–2, 129.

90 Hallé, *Éloge de la plante*, 322, 325. Burgat's phenomenological concepts of the transcendental ego (*Qu'est-ce qu'une plante?* 51), of lived experience (78), interiority, consciousness and intentionality (79) are not touched by her analysis of plant life; indeed, they are the background of the negative comparison – these are what plants do not have.

Epilogue

1 'Protista' is a semi-technical term used to group all organisms, not necessarily related, that do not fall into the categories of animals, plants and fungi.

2 Richard Owen, *Lectures on Comparative Anatomy and Physiology of the Invertebrate Animals, Delivered at the Royal College of Surgeons in 1843*, Longman, Brown, Green and Longmans, London, 1843, 374, 379. The distinction is a version of Aristotle's distinction between differences in degree (or of the 'more or less') within genera and differences in kind (across genera), where the latter can only be compared analogously. Owen was an enthusiastic reader of Aristotle, whom he still considered to be a thinker useful for science.

3 Homology is also contrasted with 'homoplasy', a wider term for the separate development of similar characters in lineages not of common ancestry, that for some includes analogy as one of its forms. See, for example, George Gaylord Simpson, *Principles of Animal Taxonomy*, Columbia University Press/ Oxford University Press, New York and London, 1961, 78. Of course, there are controversies surrounding the use of both 'homology' and 'analogy', with homology always being the more problematic concept. See Simpson, *Principles of Animal Taxonomy*, 78–88; Rolf Sattler, 'Homology – A Continuing Challenge', *Systematic Botany*, Vol. 9, No. 4 (October–December 1984), 382–94; Brian K. Hall, ed., *Homology: The*

Hierarchical Basis of Comparative Biology, Academic Press, London, 1994; Gregory R. Bock and Gail Cardew, eds., *Homology*, Wiley-Blackwell, Hoboken, NJ, 1999.

4 See Annette W. Coleman, 'Pan-eukaryote ITS2 Homologies Revealed by RNA Secondary Structure', *Nucleic Acids Research*, Vol. 35, No. 10 (15 May 2007): 3322–9.

5 Ursula Goodenough and Joseph Heitman, 'Origins of Eukaryotic Sexual Reproduction', *Cold Spring Harb Perspect Biol*, Vol. 6 (2014): 1–21. For a speculation on the last common ancestor of plants and animals, see Elliot M. Meyerowitz, 'Plants Compared to Animals: The Broadest Comparative Study of Development', *Science*, New Series, Vol. 295, No. 5559 (22 February 2002): 1482–5. Joseph Heitman has argued that LECA was most likely not sexed: 'Evolution of Sexual Reproduction: A view from the Fungal Kingdom Supports an Evolutionary Epoch with Sex before Sexes', *Fungal Biology Reviews*, Vol. 29 (2015): 108–17. This view does not seem to be contentious.

6 Margulis and Sagan, *Origins of Sex*, 2, 9, 25, 28, 56. For a discussion of viral sexuality (specifically the influenza A virus FLUAV responsible for common human influenza), see Jianping Xu, 'The Prevalence and Evolution of Sex in Microorganisms', *Genome*, Vol. 47, No. 5 (2004): 775–80.

7 This is the definition employed in Beukeboom and Perrin, *The Evolution of Sex Determination*, 3 and passim. '[M]eiotic sex is defined as a two-step process: syngamy (fusion of two haploid cells) to produce the diploid zygote; and then reduction to haploidy via meiosis, involving recombination' (3). This means that self-fertilization and parthenogenesis both count as sex but virus transmission does not.

8 Beukeboom and Perrin, *The Evolution of Sex Determination*, 3. 'Oogametes' are gametes of different sizes (so they are anisogametes) *and* different forms; basically, spermatozoa and ova. Rather surprisingly, Margulis and Sagan (*Origins of Sex*, 54) describe the donor and recipient partners in bacterial mating as male and female, respectively: 'The donor is conceptually the "male", because the genes travel from "him" to be received by "his" partner, conceptually a "female".' In *E. coli*, for example, the donor carries a replicon called the F-factor (fertility factor). In conjugation the recipient takes up the F-factor and the donor loses it; thus, in Margulis and Sagan's terms, the 'male' then becomes the 'female' and vice versa (*Origins of Sex*, 54–5). In fact, Margulis and Sagan recognize that 'male' and 'female' are 'strongly charged labels' and that the more neutral 'donor' and 'recipient' are more common (55). Roberta Bivins, 'Sex Cells: Gender and the Language of Bacterial Genetics', *Journal of the History of Biology*, Vol. 33 (2000): 113–39, is an excellent account of the discovery of bacterial sexuality and the gendering of bacteria.

9 See Stella Sandford, 'A Thousand Tiny "Sexes" – or None?', *Sluice*, December 2019. http://sluice.info/articles/sexes.html

10 Margulis and Sagan, *Origins of Sex*, 5.

11 See, for example, David L. Kirk, 'Oogamy: Inventing the Sexes', *Current Biology*, Vol. 16, No. 24 (2006): R1028–R1030, R1028; Jussi Lehtonen and Geoff A. Parker, 'Gamete Competition, Gamete Limitation, and the Evolution of the two Sexes', *Molecular Human Reproduction*, Vol. 20, No. 12 (2014): 1161–8, 1162.

12 Coulter, *The Evolution of Sex in Plants*, 102–3.

13 A suggestion that obviously, *mutatis mutandis*, owes something to Marder's 'plant thinking'.

14 Although it is not clear that 'male' and 'female', what Coulter calls 'words rather than explanations', are not also metaphysical categories, and that therefore even this characterization of anisogamy in vegetal sex is not in some sense anthropomorphic.

BIBLIOGRAPHY

Allaby, Michael. *Plant Love: The Scandalous Truth about the Sex Life of Plants*, Filbert Press, 2016.

Allaby, Michael. *A Dictionary of Plant Sciences*, fourth edition, Oxford University Press, Oxford, 2019.

Alpi, Amadeo, et al. 'Plant Neurobiology: No Brain, No Gain?', *Trends in Plant Science*, Vol. 12, No, 4 (2007): 134–5.

Alston, Charles. *A Dissertation on Botany*, Benjamin Dodds, London, 1754.

Apt Russell, Sharman. *Anatomy of a Rose: The Secret Life of Flowers*, William Heinemann, London, 2001.

Arber, Agnes. *Herbals: Their Origin and Evolution*, Lost Library, Glastonbury, undated [1912].

Arber, Agnes. 'Nehemiah Grew 1641–1712', in F. W. Oliver, ed., *Makers of British Botany: A Collection of Biographies by Living Botanists*, Cambridge University Press, Cambridge, 1913.

Arber, Agnes. 'Analogy in the History of Science', in M. F. Ashley Montagu, ed., *Studies and Essays in the History of Science and Learning, Offered in Homage to George Sarton*, Henry Schuman, NY, 1944.

Aristotle. *Physics*, trans. W. Charlton, Clarendon Press, Oxford, 1970.

Aristotle. *Generation of Animals*, trans. A. Platt, in *The Complete Works of Aristotle*, Volume II, Princeton University Press, Princeton NJ, 1984.

Aristotle. *History of Animals*, trans. D'Arcy Wentworth Thompson, in *The Complete Works of Aristotle*, Volume I, Princeton University Press, Princeton NJ, 1984.

Aristotle. *Metaphysics*, trans. W. D. Ross, in *The Complete Works of Aristotle*, Volume II, Princeton University Press, Princeton NJ, 1984.

Aristotle. 'On Youth, Old Age, Life and Death, and Respiration', trans. G. R. T. Ross, in *The Complete Works of Aristotle*, Volume I, Princeton University Press, Princeton NJ, 1984.

Aristotle. *Parts of Animals*, trans. W. Ogle, in *The Complete Works of Aristotle*, Volume I, Princeton University Press, Princeton NJ, 1984.

Aristotle. *Posterior Analytics*, trans. Jonathan Barnes, in *The Complete Works of Aristotle*, Volume I, Princeton University Press, Princeton NJ, 1984.

Aristotle. 'Sense and Sensibilia', trans. J. I. Beare, in *The Complete Works of Aristotle*, Volume I, Princeton University Press, Princeton NJ, 1984.

Aristotle. *Parts of Animals*, trans. D. M. Balme, Clarendon Press, Oxford, 1992.

Aristotle. *De Anima*, trans. Christopher Shields, Clarendon Press, Oxford, 2016.

Atran, Scott. *Cognitive Foundations of Natural History: Towards an Anthropology of Science*, Cambridge University Press, Cambridge, 1990.

Attenborough, David. *The Private Life of Plants: A Natural History of Plant Behaviour*, BBC Books, London, 1995.

Bachelard, Gaston. 'L'object de l'histoire des sciences', in *Études d'histoire et de philosophie des sciences*, Vrin, Paris, 1970.

Bacon, Francis. (Francis Lord Verulam Viscount St Albans), *Sylva Sylvarum, or a Naturall History in Ten Centuries*, ed. W. Rawley, William Lee, London, 1639.

Bailey, L. H. 'On the Untechnical Terminology of the Sex-Relation in Plants', *Science*, Friday 5 June 1896: 825–7.

Baker, Henry. 'The Discovery of a Perfect Plant in Semine', *Philosophical Transactions*, Vol. 41 (1739–41): 448–55.

Baldwin, Ian T. and Jack C. Schultz. 'Rapid Changes in Tree Leaf Chemistry Induced by Damage: Evidence for Communication between Plants', *Science*, Vol. 221, No. 4607 (15 July 1983): 227–9.

Baldwin, Ian T., André Kessler, and Rayko Halitschke. 'Volatile Signalling in Plant-plant Herbivore Interactions: What Is Real?', *Current Opinion in Plant Biology*, Vol. 5 (2002): 1–4.

Baldwin, Ian T., Rayko Halitschke, and Anja Paschold. 'Volatile Signalling in Plant-plant Interactions: "Talking Trees" in the Genomics Era', *Science*, Vol. 311 (2006): 812–15.

Baluška, František and Stefano Mancuso. 'Vision in Plants via Plant-Specific Ocelli?', *Trends in Plant Science*, September, Vol. 21, No. 9 (2016): 727–30.

Barker, M. 'Putting Thought in Accordance with Things: The Demise of Animal-Based Analogies for Plant Functions', *Science & Education*, Vol. 11, No. 3 (2002): 293–304.

Barrett, Spencer C. H. and Deborah Charlesworth. 'David Graham Lloyd, 20 June 1937–30 May 2006', *Biographical Memoirs of Fellows of the Royal Society*, Vol. 53 (2007): 203–21.

Barrett, Spencer C. H. and Laurence D. Harder. 'David G. Lloyd and the Evolution of Floral Biology: From Natural History to Strategic Analysis', in Harder and Barrett, eds., *Ecology and Evolution of Flowers*, Oxford University Press, Oxford, 2006, 1–21.

Beiler, Kevin J., Daniel M. Durall, Suzanne W. Simard, Sheri A. Mazwell, and Annette M. Kretzer. 'Architecture of the Wood-Wide Web: *Rhizopogon* spp. Genets Link Multiple Douglas-fir Cohorts', *New Phytologist*, Vol. 185 (2010): 543–53.

Bell, Peter R. 'The Alternation of Generations', in J. A. Callow, ed., *Advances in Botanical Research*, Volume 16, Academic Press, London, 1989, 55–93.

Bernasconi, Paul and Lincoln Taiz. 'Sebastien Vaillant's 1717 Lecture on the Structure and Function of Flowers', *HUNTIA: A Journal of Botanical History*, Vol. 11, No. 2 (2002): 97–128.

Beukeboom, Leo W. and Nicolas Perrin, *The Evolution of Sex Determination*, Oxford University Press, Oxford, 2014.

Bingham, Marcus A. and Suzanne W. Simard. 'Ectomycorrhizal Networks of *Pseudotsuga menziesii* var. *glauca* [Rocky Mountain Douglas Fir]

Trees Facilitate Establishment of Conspecific Seedlings under Drought', *Ecosystems*, Vol. 15 (2012): 188–99.

Bivins, Roberta. 'Sex Cells: Gender and the Language of Bacterial Genetics', *Journal of the History of Biology*, Vol. 33 (2000): 113–39.

Black, Max. 'Metaphor', *Proceedings of the Aristotelian Society*, New Series, Vol. 55 (1954–5): 273–94.

Black, Max. *Models and Metaphors: Studies in Language and Philosophy*, Cornell University Press, Ithaca NY, 1962.

Blair, Patrick. *Botanick Essays in Two Parts*, London, Printed by Willian and John Innys … at the Prince's Arms, 1720.

Blunt, Wilfrid. *The Compleat Naturalist: A Life of Linnaeus*, Frances Lincoln, London, 2001 [1971].

Bock, Gregory R. and Gail Cardew, eds. *Homology*, Wiley-Blackwell, Hoboken, NJ, 1999.

Bonner, John Tyler. 'Analogies in Biology', *Synthese*, Vol. 15 (1963): 275–279.

Borges, Renee. 'Gender in Plants', *Resonance*, Vol. 3, No. 11 (November 1998): 30–9.

Bower, F. O. *Botany of the Living Plant*, Macmillan & Co, London, fourth edition, 1947 [1919].

Bradley, Richard. *New Improvements of Planting and Gardening, Both Philosophical and Practical, Explaining the Motion of the Sap and Generation of Plants & …*, W. Mears, London, 1719 [1717].

Bradley, Richard. 'The Analogy between Plants and Animals, Drawn from the Difference of Their Sexes', in *A Philosophical Account of the Works of Nature*, W. Mears, London, 1721.

Brenner, Eric D., Rainer Stahlberg, Stefano Mancuso, Jorge Vivanco, František Baluška, and Elizabeth Van Volkenburgh. 'Plant Neurobiology: An Integrated View of Plant Signalling', *Trends in Plant Science*, Vol. 11, No. 8 (2006): 413–19.

Bremekamp, C. E. B. 'A Re-Examination of Cesalpino's Classification', *Acta botanica neerlandica*, Vol. 1, No. 4 (1953): 580–93.

Bretin-Chabrol, Marine et Claudine Leduc. 'La botanique antique et la problématique du genre', *Clio. Femmes, Genre, Histoire*, Vol. 29 (2009): 205–23.

Brown, Robert. 'Observations on the Structure and Affinities of the More Remarkable Plants Collected by… Walter Oudney etc', London, 1826, reprinted in Brown, *Miscellaneous Botanical Works*, Volume I, Ray Society, London, 1866.

Browne, Janet. 'Botany for Gentlemen: Erasmus Darwin and "The Loves of the Plants"', *Isis*, Vol. 80, No. 4 (1989): 593–621.

Burgat, Florence. *Qu'est-ce qu'une plante? Essai sur la vie végétale*, Seuil, Paris, 2020.

Calvo, Paco. 'The Philosophy of Plant Neurobiology', *Synthese*, Vol. 193 (2016): 1323–43.

Calvo, Paco and František Baluška. 'Conditions for Minimal Intelligence across Eukaryota: A Cognitive Science Perspective', *Frontiers in Psychology*, 3 September 2015.

Calvo, Paco, Monica Gagliano, Gustavo M. Souza, and Anthony Trewavas. 'Plants Are Intelligent, Here's How', *Annals of Botany*, Vol. 125 (2020): 11–28.

Camerarii, Rudolphi Jacobi. [Camerarius]. *De sexu plantarum epistola*, Tubingae, 1694.

Camerarius, Rudolf Jakob. *Das Geschlecht der Pflanzen*, trans. M. Möbius and Wilhelm Engel, Wilhelm Engel, Leipzig, 1899.

Campbell, Neil A., et al. *Biology*, eighth edition, Pearson Benjamin Cummings, San Francisco CA, 2008.

Canguilhem, Georges. 'Modèles et analogies dans la découverte en biologie', in *Études d'histoire et de philosophie des sciences*, Vrin, Paris, 1970 [1968].

Carey, Nessa. *The Epigenetics Revolution: How Modern Biology Is Rewriting Our Understanding of Genetics, Disease and Inheritance*, Icon, London, 2012.

Cesalpini Arentini, Andreae [Cesalpino]. *Peripateticarum Quaestionem*, Venetiis, Venetiis, 1571.

Cesalpini Arentini, Andreae [Cesalpino]. *De plantis libri XVI*, Apud Georgium Marescottum, Florentiae, 1583.

Chamberlain, Charles J. 'Alternation of Generations in Animals from a Botanical Standpoint', *Science*, Vol. 22, No. 555 (18 August 1905): 137–44.

Chamowitz, Daniel. *What a Plant Knows: A Field Guide to the Senses of Your Garden and Beyond*, Oneworld, London, 2012.

Cheung, Tobias. 'From the Organism of a Body to the Body of an Organism: Occurrence and Meaning of the Word "Organism" from the Seventeenth to the Nineteenth Centuries', *British Journal for the History of Science*, Vol. 39, No. 3 (September 2006): 319–39.

Coccia, Emanuele. *The Life of Plants: A Metaphysics of Mixture*, trans. Dylan J. Montanari, Polity, Cambridge, 2019 [2017].

Coleman, Annette W. 'Pan-eukaryote ITS2 Homologies Revealed by RNA Secondary Structure', *Nucleic Acids Research*, Vol. 35, No. 10 (15 May 2007): 3322–9.

Corner, E. J. H. *The Life of Plants*, University of Chicago Press, Chicago and London, 2002 [1964].

Coulter, John Merle. *Plant Studies: An Elementary Botany*, D. Appleton and Company, New York, 1901.

Coulter, John Merle. *A Text-Book of Botany for Secondary Schools*, D. Appleton and Company, New York, 1910.

Coulter, John Merle. *The Evolution of Sex in Plants*, University of Chicago Press, Chicago IL, 1914.

Coulter, John Merle and Charles J. Chamberlain. *Morphology of Gymnosperms*, University of Chicago Press, Chicago, 1910 [1901].

Coulter, John Merle and Charles J. Chamberlain. *Morphology of Angiosperms*, D. Appleton and Company, New York, 1903.

Crepy, María A. and Jorge J. Casal. 'Photoreceptor-mediated Kin Recognition in Plants', *New Phytologist*, Vol. 205 (2015): 329–38.

Darwin, Charles. *The Different Forms of Flowers on Plants of the Same Species*, second edition, D. Appleton, New York, 1899 [1877].

Darwin, Charles, assisted by Francis Darwin. *The Power of Movement in Plants*, John Murray, London, 1880.

Darwin, Erasmus. *Zoonomia, or, the Laws of Organic Life*, J. Johnson, London 1794.

Darwin, Erasmus. The Botanic Garden. A Poem, Part II, in *The Loves of the Plants*, J. Johnson, London, 1799 [1789].

Darwin, Francis. 'The Analogies of Plant and Animal Life', *Nature*, Part I, 14 March; Part II, 21 March 1878.

Daugey, Fleur. *Les plantes ont-elles un sexe? Histoire d'une découverte*, Éditions Ulmer, Paris, 2015.

Daugey, Fleur. *L'intelligence des plantes: Les découverts qui révolutionnent notre comprehension du monde végétal*, Les Éditions Ulmer, Paris, 2018.

Delaporte, François. *Nature's Second Kingdom: Explorations of Vegetality in the Eighteenth Century*, trans. Arthur Goldhammer, MIT Press, Cambridge MA, 1982 [1979].

Delph, Lynda F. and David G. Lloyd. 'Environmental and Genetic Control of Gender in the Dimorphic Shrub Hebe subalpina', *Evolution*, Vol. 45, No. 8 (December 1991): 1957–64.

Derrida, Jacques. 'White Mythology' (1971), in *Margins of Philosophy*, Alan Bass, trans., Harvester Wheatsheaf, New York etc, 1982 [1972].

Descola, Philippe. *Beyond Nature and Culture*, trans. Janet Lloyd, University of Chicago Press, Chicago, 2013 [2005].

Dorolle, Maurice. 'Introduction', in Césalpin, *Questions Péripatéticiennes*, trans. Maurice Dorolle, Librarie Félix Alcan, Paris, 1929.

Drossaart Luflos, H. J. and E. L. J. Poortman, eds. *Nicolaus Damacenus, De Plantis: Five Translations*, North Holland Publishing Company, Amsterdam, 1989.

Ducker, Sophie C. and R. Bruce Knox. 'Pollen and Pollination: A Historical Review', *Taxon*, Vol. 34, No. 3 (August 1985): 401–19.

Dyachenko, Alexander P. 'Botanical Terminology: New Twists or Tradition?', *Skvortsovia*, Vol. 3, No. 3 (2017): 122–9.

Enger, Eldon D., Frederick C. Ross, and David B. Bailey. *Concepts in Biology*, twelth edition, McGraw Hill, Boston etc., 2007.

Erikson, Gunnar. 'Linnaeus the Botanist', in Tore Frängsmyr, ed., *Linnaeus: The Man and His Work*, University of California Press, Los Angeles/ London, 1983.

Fara, Patricia. *Sex, Botany and Empire: The Story of Carl Linnaeus and Joseph Banks*, Icon Books, Duxford, Cambridge, 2003.

Farley, John. *Gametes and Spores: Ideas about Sexual Reproduction 1750–1914*, Johns Hopkins University Press, Baltimore and London, 1982.

Fellner, Stephan. *Albertus als Botaniker*, Alfred Hölder, Wien, 1881.

Fox Keller, Evelyn. *Making Sense of Life: Explaining Biological Development with Models, Metaphors and Machines*, Harvard University Press, Cambridge MA, 2002.

Foxall, Lin. 'Natural Sex: The Attribution of Sex and Gender to Plants in Ancient Greece', in Lin Foxhall and John Salmon, eds., *Thinking Men: Masculinity and its Self-Representation in the Classical Tradition*, Routledge, London, 1998.

Frankel, R. and E. Galun. *Pollination Mechanisms, Reproduction and Plant Breeding*, Springer-Verlag, Berlin, 1977.

Funk, Holger. 'Adam Zalužanský's "De sexu plantarum" (1592): An Early Pioneering Chapter on Plant Sexuality', *Archives of Natural History*, Vol. 40, No. 2 (2013): 244–56.

Gagliano, Monica. 'Green Symphonies: A Call for Studies on Acoustic Communication in Plants', *Behavioural Ecology*, Vol. 24, No. 4 (July–August 2013): 789–96.

Gagliano, Monica and Martial Depczynski. 'Tuned In: Plant Roots Use Sound to Locate Water', *Oecologica*, Vol. 184, No. 1 (2017): 151–60.

Gagliano, Monica, Stefano Mancuso and Daniel Robert. 'Towards Understanding Plant Bioacoustics', *Trends in Plant Science*, Vol. 17, No. 6 (June 2012): 323–5.

Garrett, Brian. 'Vitalism and Teleology in the Natural Philosophy of Nehemiah Grew (1641–1712)', *BJHS*, Vol. 36, No. 1 (March 2003): 63–81.

Gelber, Jessica. 'Females in Aristotle's Embryology', in Andrea Falcon and David Lefebvre, eds., *Aristotle's Generation of Animals: A Critical Guide*, Cambridge University Press, Cambridge, 2018.

Geoffroy, Claude Joseph. 'Observations sur la structure et l'usage des principals parties des fleurs', *Mémoires de mathématiques et de physique de l'Académie royale des sciences* (14 November 1711): 207–31.

George, Sam. *Botany, Sexuality and Women's Writing 1760–1830: From Modest Shoot to Forward Plant*, Manchester University Press, Manchester, 2007.

Gianoli, Ernesto. 'Eyes in the Chameleon Vine?', *Trends in Plants Science*, Vol. 22, No. 1 (1 January 2017): 4–5.

Gianoli, Ernesto and Fernando Carrasco-Urra. 'Leaf Mimicry in a Climbing Plant Protects against Herbivory', *Current Biology*, Vol. 24 (5 May 2014): 984–7.

Gingras, Yves and Alexandre Guay. 'The Uses of Analogies in Seventeenth and Eighteenth Century Science', *Perspectives on Science*, Vol. 19, No. 2 (2011): 154–91.

Gmelin, Johann Georg. *Sermo Academicus de Novorum Vegetabilium Post Creationem Divinam Exortu*, Literis Erhardditianus, Tubinngae, 1749.

Goodenough, Ursula and Joseph Heitman. 'Origins of Eukaryotic Sexual Reproduction', *Cold Spring Harb Perspect Biol*, Vol. 6 (2014): 1–21.

Gorzelak, Monika A., Amanda K. Asay, Brian J. Pickles and Suzanne Simard. 'Inter-Plant Communication through Mycorrhizal Networks Mediates Complex Adaptive Behaviour in Plant Communities', *AoB Plants*, Vol. 7 (2015): 1–13.

Gotthelf, A. 'Introduction' to Aristotle, *History of Animals*, Loeb, Harvard University Press, Princeton, NJ, 1991.

Grew, Nehemiah. *The Anatomy of Plants*, W. Rawlins, 1682.

Grew, Nehemiah. *Cosmologia Sacra, or a Discourse of the Universe as It Is the Creature and Kingdom of God, etc.*, W. Rogers, S. Smith and B. Walford, London, 1701.

Hagberg, Knut. *Carl Linnaeus*, trans. Alan Barr, Jonathan Cape, London, 1952.

Haig, David. 'Homologous Versus Antithetic Alternation of Generations and the Origin of Sporophytes', *Botanical Review*, Vol. 74, No. 3 (September 2008): 395–418.

Haig, David and Mark Westoby. 'Inclusive Fitness, Seed Resources, and Maternal Care', in Jon Lovett Doust and Lesley Lovett Doust, eds., *Plant Reproductive Ecology: Patterns and Strategies*, Oxford University Press, Oxford, 1988.

Hall, Brian K., ed. *Homology: The Hierarchical Basis of Comparative Biology*, Academic Press, London, 1994.

Hall, Matthew. *Plants as Persons: A Philosophical Botany*, SUNY Press, New York, 2011.

Hallé, Francis. *Tropical Trees and Forests: An Architectural Analysis*, Springer, New York, 1978.

Hallé, Francis. *Éloge de la plante: Pour une nouvelle biologie*, Seuil, Paris, 1999.

Hallé, Francis, *La vie des arbres*, Bayard, Montrogue, 2011.

Hallyn, Fernand, ed. *Metaphor and Analogy in the Sciences*, Kluwer, Dordrecht, 2000.

Harper, John L. *Population Biology of Plants*, Academic Press, London etc, 1977.

Harshburger, John W. 'James Logan, an Early Contributor to the Doctrine of Sex in Plants', *Botanical Gazette*, Vol. 19, No. 8 (August 1894): 307–12.

Harvey-Gibson, R. J. *Outlines of the History of Botany*, A&C Black, London, 1919.

Hawks, Ellison. *Pioneers of Plant Study*, Books for Libraries Press, Freeport New York, 1928.

Heidegger, Martin. *Being and Time*, trans. John Macquarrie and Edward Robinson, Basil Blackwell, Oxford, 1962 [1927].

Heitman, Joseph. 'Evolution of Sexual Reproduction: A View from the Fungal Kingdom Supports an Evolutionary Epoch with Sex before Sexes', *Fungal Biology Reviews*, Vol. 29 (2015): 108–17.

Heslop-Harrison, J. S. 'The Angiosperm Stigma', in M. Cresti and A. Tiezzi, eds., *Sexual Plant Reproduction*, Springer-Verlag, Berlin etc., 1992, 59–68.

Hesse, Mary B. *Models and Analogies in Science*, University of Notre Dame Press, Indiana, 1966.

Hesse, Mary B. 'Tropical Talk: The Myth of the Literal', Richard Rorty and Mary Hesse 'Unfamiliar Noises': II, *Proceedings of the Aristotelian Society, Supplementary Volumes*, Vol. 61 (1987): 283–311.

Hesse, Mary B. 'Models, Metaphors and Truth', in F.R. Ankersmit and J.J.A. Mooij, eds., *Knowledge and Language*, Vol. III, Kluwer, Dordrecht, 1993: 49–66.

Hiernaux, Quentin. 'Pourquoi et comment philosopher sur le végétal?', in Hiernaux and Timmermans, eds., *Philosophie du végétal* Vrin, Paris, 2019.

Hiernaux, Quentin and Benoît Timmermans, eds. *Philosophie du végétal*, Vrin, Paris, 2019.

Hofmeister, William *On the Germination, Development, and Fructification of the Higher Cryptogamia and on the Fructification of the Coniferae*, trans. Frederick Curry, Ray Society, London, 1862 [1851].

Hooper, Robert. *Observations on the Structure and Economy of Plants, to Which Is Added the Analogy between the Animal and the Vegetable Kingdom*, Oxford, 1797.

Ingensiep, Hans Werner. 'The History of the Plant Embryo. Terminology and Visualization from Ancient until Modern Times', *Hist. Phil. Life Sci.*, Vol. 26 (2004): 309–31.

Irigaray, Luce and Michael Marder. *Through Vegetal Being: Two Philosophical Perspectives*, Columbia University Press, NY, 2016.

Jerling, Lenn. 'Are Plants and Animals Alike? A Note on Evolutionary Plant Population Ecology', *Oikos*, Vol. 45, No. 1 (August 1985): 150–3.

Johansen, Donald Alexander. *Plant Embryology: Embryogeny of the Spermatophyta*, Chronica Botanica Company, Waltham, Mass., 1950.

Jones, Roy W. 'Alternation of Generations as Expressed in Plants and Animals', *Bios*, Vol. 8, No. 1 (March 1937): 19–29.

Jussieu, Antoine de. *Du rapport des plantes avec les animaux tiré de la différence de leurs sexes*, 1721.

Kant, Immanuel. *Critique of Pure Reason*, trans. Paul Guyer and Allan W. Wood, Cambridge University Press, Cambridge, 1997 [1781/1787].

Karban, Richard. 'Plant Behaviour and Communication', *Ecology Letters*, Vol. 11 (2008): 727–39.

Karban, Richard. *Plant Sensing and Communication*, University of Chicago Press, Chicago, IL, 2015.

Kassinger, Ruth. *A Garden of Marvels: How We Discovered that Flowers Have Sex, Leaves Eat Air, and Other Secrets of Plants*, William Morrow, New York, 2012.

Kaufmann, H., H. Kirsch, T. Wemmer, A. Peil, F. Lottspeich, H. Ulrig, F Salamini, and R. Thompson. 'Sporophytic and Gametophytic Self-Incompatibility', in M. Cresti and A. Tiezzi, eds., *Sexual Plant Reproduction*, Springer-Verlag, Berlin etc., 1992, 115–25.

Kew Botanical Gardens. *Flora: Inside the Secret World of Plants*, Dorling Kindersley, London, 2018.

Kirk, David L. 'Oogamy: Inventing the Sexes', *Current Biology*, Vol. 16, No. 24 (2006): R1028–R1030.

Kohn, Eduardo. *How Forest Think: Toward an Anthropology beyond the Human*, University of California Press, Berkeley, 2013.

Kutschera, Ulrich and Karl J. Niklas. 'Julius Sachs (1868): The Father of Plant Physiology', *American Journal of Botany*, Vol. 105, No. 4: 656–66.

Larson, James L. 'Linnaeus and the Natural Method', *Isis*, Vol. 58, No. 3 (Autumn 1967): 304–20.

Larson, James L. 'Linnean Analogy', *Scandanavian Studies*, Vol. 40, No. 4 (November 1968): 294–302.

Leatherdale, W. H. *The Role of Analogy, Model and Metaphor in Science*, North-Holland Publishing Company, Amsterdam/Oxford, 1974.

Le Doeuff, Michèle. *The Philosophical Imaginary*, trans. Colin Gordon, Continuum, London and New York, 2002 [1980].

Lee, James. *An Introduction to Botany Containing Explanations of the Theory of that Science and an Interpretation of Its Technical Terms, Extracted from the Works of Dr Linnaeus ....*, Printed for J. and R. Tonson, in the Strand, London, 1760.

Le Guin, Ursula K. 'Vaster Than Empires and More Slow' [1971], in *The Wind's Twelve Quarters*, Victor Gollancz, London, 1989.

Lehtonen, Jussi and Geoff A. Parker. 'Gamete Competition, Gamete Limitation, and the Evolution of the two Sexes', *Molecular Human Reproduction*, Vol. 20, No. 12 (2014): 1161-8.

Lindroth, Sten. 'The Two Faces of Linnaeus', in Tore Frängsmyr, ed., *Linnaeus: The Man and His Work*, University of California Press, Los Angeles/ London, 1983.

Linnaei, Caroli [Carl Linnaeus]. *Systema Naturae*, Theodorum Haak, Lugduni Batavorum [Leiden], 1735.

Linnaei, Caroli [Carl Linnaeus]. *Fundamenta Botanica*, Salomonem Schouten, Amstelodami, 1736.

Linnaei, Caroli [Carl Linnaeus]. *Philosophia botanica*, Godofr. Kiesewetter, Stockholm, 1751.

Linnaeus, Carl. *Prelude to the Betrothal of Plants*, trans. Clare James, Uppsala Universiteitsbibliotek, Uppsala, 2007 [1729].

Linnaeus, Carl. *Philosophia Botanica*, trans. Stephen Freer, Oxford University Press, Oxford, 2003 [1751].

Linnaeus, Carl. *A System of Vegetables*, trans. Erasmus Darwin and others, from the 13th edition of the Systema Vegetabilium… By a Botanical Society at Lichfield, Lichfield 1782/83.

Linnaeus, Carl. *Dissertation on the Sexes of Plants*, trans. James Edward Smith, George Nicol, London, 1786.

Linnaeus, Carl. *Praelectiones in Ordines Naturales Plantarum*, ed. Paulus Diet, Giseke, Impr, Benj. Gottl. Hoffmanni, 1792.

Lloyd, David G. 'Breeding Systems in Cotula L. (Compositae, Anthemideae). I. The Array of Monoclinous and Diclinous Systems', *The New Phytologist*, Vol. 71, No. 6 (November 1972): 1181-94.

Lloyd, David G. 'Female-Predominant Sex Ratios in Angiosperms', *Heredity*, Vol. 32, No. 1 (1974): 35-44.

Lloyd, David G. 'Theoretical Sex Ratios of Dioecious and Gynodioecious Angiosperms', *Heredity*, Vol. 32, No. 1 (1974): 11-34.

Lloyd, David G. 'The Transmission of Genes via Pollen and Ovules in Gynodioecious Angiosperms', *Theoretical Population Biology*, Vol. 9, No. 3 (1976): 299-316.

Lloyd, David G. 'Parental Strategies of Angiosperms', *New Zealand Journal of Botany*, Vol. 17 (1979): 595-606.

Lloyd, David G. 'Sexual Strategies in Plants I: An Hypothesis of Serial Adjustment of Maternal Investment During One Reproductive Season', *The New Phytologist*, Vol. 86 (1980): 69-79.

Lloyd, David G. 'Sexual Strategies in Plants III: A Quantitative Method for Describing the Gender of Plants', *New Zealand Journal of Botany*, Vol. 18, No. 1 (1980): 103-8.

Lloyd, David G. 'The Distribution of Gender in Four Angiosperm Species Illustrating Two Evolutionary Pathways to Dioecy', *Evolution*, Vol. 34, No. 1 (January 1980): 123-34.

Lloyd, David G. and K. S. Bawa. 'Modification of the Gender of Seed Plants in Varying Conditions', *Evolutionary Biology*, Vol. 17 (1984): 255–338.

Lloyd, G. E. R. *Analogical Investigations: Historical and Cross-cultural Perspectives on Human Reasoning*, Cambridge University Press, Cambridge, 2015.

Logan, James. 'Some Experiments Concerning the Impregnation of the Seeds of Plants', *Philosophical Transactions*, Vol. 39, No. 4, (31 December 1735): 192–5.

Magni, Alberti. *De Vegetablibus Libri VII*, Editionem Criticam ab Ernesto Meyero Coeptam, Berolini, 1867.

Malpighi, Marcelli. *Anatome Plantarum*, Regiae Societati, Londini, 1675.

Mamut, Jannathan, Ying-Ze Xiong, Dun-Yan Tan, and Shuang-Quan Huang. 'Pistillate Flowers Experience More Pollen Limitation and Less Geitonogamy than Perfect Flowers in a Gynomonoecious Herb', *The New Phytologist*, Vol. 201, No. 2 (January 2014): 670–7.

Mancuso, Stefano. *Brilliant Green: The Surprising History and Science of Plant Intelligence*, trans. Joan Benham, Island Press, Washington, 2015 [2013].

Mancuso, Stefano. *The Revolutionary Genius of Plants: A New Understanding of Plant Behaviour and Intelligence*, trans. Vanessa Di Stefano, Atria Books, NY etc, 2018 [2017].

Marder, Michael. *Plant Thinking: A Philosophy of Vegetal Life*, Columbia University Press, 2013.

Marder, Michael. *Grafts: Writings on Plants*, Univocal, Minneapolis, 2016.

Margulis, Lynn and Dorion Sagan. *Origins of Sex: Three Billion Years of Genetic Recombination*, Yale University Press, New Haven, 1986.

Mascarenhas, P. 'The Male Gametophyte of Flowering Plants', *The Plant Cell*, Vol. 1 (July 1989): 657–64.

Mauseth, James D. *Botany: An Introduction to Plant Biology*, fifth edition, Jones & Bartlett, Burlington, MA, 2014.

Mayhew, Robert. *The Female in Aristotle's Biology: Reason or Rationalization*, University of Chicago Press, Chicago and London, 2004.

McClintock, Barbara. 'The Origin and Behaviour of Mutable Loci in Maize', *Proceedings of the National Academy of Sciences of the United States of America*, Vol. 36, No. 6 (1950): 344–55.

McGillivray, W. *Lives of Eminent Zoologists from Aristotle to Linnaeus*, Oliver & Boyd, London, 1834.

Meeker, Natania and Antónia Szabari. *Radical Botany: Plants and Speculative Fiction*, Fordham University Press, New York, 2020.

Meeuses, Bastiaan and Sean Morris. *The Sex Life of Flowers*, Faber and Faber, London/Boston, 1984.

Mettrie, Julien Offray de la. *L'Homme plante*, in *Oeuvres philosophiques*, Tome Second, Charles Tutot, Paris, 1796 [1747].

Meyerowitz, Elliot M. 'Plants Compared to Animals: The Broadest Comparative Study of Development', *Science*, New Series, Vol. 295, No. 5559 (22 February 2002): 1482–5.

Meyers, Lauren Ancel and Donald A. Levin. 'On the Abundance of Polyploids in Flowering Plants', *Evolution*, Vol. 60, No. 6 (2006): 1198–206.

Miall, L. C. *The Early Naturalists: Their Lives and Work (1530–1789)*, Macmillan & Co, New York, 1912.

Mill, John Stuart. *A System of Logic, Ratiocinative and Inductive*, Harper & Brothers, New York, 1882.

Miller, Elaine P. *The Vegetative Soul: From Philosophy of Nature to Subjectivity in the Feminine*, SUNY Press, New York, 2002.

Miller, Philip. *The Gardener's Dictionary*, J. Rivington, London, 1759 [1731].

Moore, John and Lindley Thomas. *The Treasury of Botany: A Popular Dictionary of the Vegetable Kingdom; with Which Is Incorporated a Glossary of Botanical Terms*, Longmans, Green and Co., London, 1866.

Morland, Samuel. 'Some New Observations upon the Parts and Use of the Flower in Plants', *Philosophical Transactions*, Vol. 23, No. 287 (01 January 1703): 1474–9.

Morsink, Johannes. *Aristotle on the Generation of Animals: A Philosophical Study*, University Press of America, Lanham NY and London, 1982.

Morton, A. G. *History of Botanical Science: An Account of the Development of Botany from Ancient Times to the Present Day*, Academic Press, London etc, 1981.

Morton, A. G. 'Marginalia to Cesalpino's Work on Botany', *Archives of Natural History*, Vol. 10, No. 1 (1981): 31–6.

Morton, A. G. 'A Letter of Andrea Cesalpino', *Archives of Natural History*, Vol. 14, No. 2 (1987): 169–173.

Mueller, Ferdinand Von. 'Considerations of Phytographic Expressions and Arrangements', *Journal and Proceedings of the Royal Society of New South Wales*, Vol. 22 (1888): 187–204.

Murray, Brian G. 'Floral Biology and Self-Incompatibility in Linum', *Botanical Gazette*, Vol. 147, No. 3 (September 1986): 327–33.

Neal, P. R. and G. J. Anderson. 'Are "Mating Systems" "Breeding Systems" of Inconsistent and Confusing Terminology in Plant Reproductive Biology? Or Is It the Other Way Around?', *Plant Systems and Evolution*, 250 (2005): 173–85.

Negbi, Moshe. 'Male and Female in Theophrastus's Botanical Works', *Journal of the History of Biology*, Vol. 28, No. 2 (Summer 1995): 317–32.

Nielsen, Karen M. 'The Private Parts of Animals: Aristotle on the Teleology of Sexual Difference', *Phronesis* 53 (2008): 373–405.

Ogilvie, Brian W. *The Science of Describing: Natural History in Renaissance Europe*, University of Chicago Press, Chicago IL, 2006.

Olson, Harry F. *Dynamical Analogies*, Chapman & Hall, London, 1943.

Owen, Richard. *Lectures on Comparative Anatomy and Physiology of the Invertebrate Animals, Delivered at the Royal College of Surgeons in 1843*, Longman, Brown, Green and Longmans, London, 1843.

Pagel, Walter. 'Cesalpino's Peripatetic Questions', *History of Science*, Vol. 13, No. 2 (1973): 130–8.

Paik, Inyup and Enamul Huq. 'Plant Photoreceptors: Multi-functional Sensory Proteins and Their Signalling Networks', *Seminars in Cell & Developmental Biology*, 92 (2019): 114–21.

Pollan, Michael. *The Botany of Desire: A Plant's Eye-View of the World*, Bloomsbury, London, 2001.

Pollan, Michael. 'The Intelligent Plant', *The New Yorker*, 23 and 30 December 2013.

Polwhele, Richard. *The Unsex'd Females: A Poem*, Cadell and Davis, London, 1798.

Powers, Richard. *The Overstory*, Vintage, London, 2018.

Primack, Richard B. and David G. Lloyd. 'Sexual Strategies in Plants IV. The Distributions of Gender in Two Monomorphic Shrub Populations', *New Zealand Journal of Botany*, 18, No. 1 (1980): 109–14.

Proctor, Michael, Peter Yeo, and Andrew Lack. *The Natural History of Pollination*, HarperCollins, London, 1996.

Pulteney, Richard. *A General View of the Writings of Linnaeus, to Which Is Annexed The Diary of Linnaeus, Written by Himself*, J. Mawman, London, 1805.

Rackham, Oliver. *Woodlands*, William Collins, London, 2015 [2006].

Ramsbottom, John. 'S. Vaillant's Discours sur la structure des fleurs', *Soc. Bibiogr. Nat. Hist*, Vol. 4 (1718): 194–6.

Raven, Charles E. *John Ray: Naturalist. His Life and Works*, second edition, Cambridge University Press, Cambridge etc, 1950 [1942].

Ray, John. *Historia Plantarum Generalis*, Samuelis Smith, London, 1643.

Ray, John. *Historia Plantarum Generalis*, trans. Margaret Lazenby, Volume 3, 1072. [Newcastle University thesis, 1995, available online.]

Ray, John. *The Wisdom of God, Manifested in the Works of Creation*, Facsimiliae of 1826 edition, Scion Publishing, Bloxham, Oxon, 2005.

Reeds, Karen. 'Albert on the Natural Philosophy of Plant Life', in James A. Weisheipl, ed., *Albertus Magnus and the Sciences: Commemorative Essays, 1980*, Pontifical Institute of Medieval Studies, Toronto, 1980.

Reynolds Green, J. *A History of Botany in the United Kingdom from the Earliest Times to the End of the 19th Century*, J.M. Dent, London & Toronto, 1914.

Riddle, John M. *Dioscorides on Pharmacy and Medicine*, University of Texas Press, Austin, 1985.

Ritterbush, Philip C. *Overtures to Biology: The Speculations of Eighteenth-Century Naturalists*, Yale University Press, New Haven & London, 1964.

Robertson, Jennifer. 'Sexy Rice: Plant Gender, Farm Manuals and Grass-Roots Nativism', *Monumenta Nipponica*, Vol. 39, No. 3 (Autumn 1984): 233–60.

Rorty, Richard. 'Hesse and Davidson on Metaphor', Richard Rorty and Mary Hesse. 'Unfamiliar Noises': I, *Proceedings of the Aristotelian Society, Supplementary Volumes*, Vol. 61 (1987): 283–311.

Sachs, Julius von. *Lehrbuch der Botanik, Nach dem Gegenwärtigen Stand der Wissenschaft*, fourth edition, Wilhelm Engelmann, Leipzig, 1874 [1868].

Sachs, Julius von. *History of Botany (1530–1860)*, trans. Henry E. F. Garnsey, Clarendon Press, Oxford, 1906 [1871–2].

Sakai, Ann K. and Stephen G. Weller. 'Gender and Sexual Dimorphism in Flowering Plants: A Review of Terminology, Biogeographic Patterns, Ecological Correlates and Phylogenetic Approaches', in M. A. Geber, T. E.

Dawson, and L. F. Delph, eds., *Gender and Sexual Dimorphism in Flowering Plants*, Springer, Berlin, 1999.

Sandford, Stella. 'From Aristotle to Contemporary Biological Classification: What Kind of Category Is "Sex"?', *Redescriptions: Political Thought, Conceptual History and Feminist Theory*, Vol. 22, No. 1 (2019): 4–17.

Sandford, Stella. 'A Thousand Tiny "Sexes" – or None?', *Sluice*, December 2019.

Sattler, Rolf. 'Homology – A Continuing Challenge', *Systematic Botany*, Vol. 9, No. 4 (October–December 1984): 382–94.

Schiebinger, Londa. *Nature's Body: Gender in the Making of Modern Science*, Beacon Press, Boston, 1993.

Schiebinger, Londa. 'Gender and Natural History', in N. Jardine, J. A. Secord, and E. C. Spary, eds., *Cultures of Natural History*, Cambridge University Press, Cambridge, 1996.

Serra, Carlo et al. 'Historical Controversies about the Thalamus: From Etymology to Function', *Journal of Neurosurgery* (Neurosurg Focus), Vol. 47, No. 3 (2019): E13.

Sheldrake, Merlin. *Entangled Life: How Fungi Make Our Worlds, Change Our Minds, and Shape Our Futures*, The Bodley Head, London, 2020.

Simard, Suzanne W. 'Mycorrhizal Networks and Complex Systems: Contributions of Soil Ecology Science to Managing Climate Change Effects in Forested Ecosystems', *Can. J. Soil Sci.*, Vol. 89 (2009): 369–82.

Simard, Suzanne W. 'Mycorrhizal Networks Facilitate Tree Communication, Learning and Memory', in F. Baluška, M. Gagliano and G. Witzany, eds., *Memory and Learning in Plants*, Springer, Cham, Switzerland, 2018, 191–213.

Simard, Suzanne W. *Finding the Mother Tree: Uncovering the Wisdom and Intelligence of the Forest*, Allen Lane, London, 2021.

Simard, Suzanne W., David A. Perry, Melanie D. Jones, David D. Myrold, Daniel M. Durall, and Randy Molina. 'Net Transfer of Carbon Between Ectomycorrhizal Tree Species in the Field', *Nature*, Vol. 388, No. 7 (August 1997): 579–82.

Simard, Suzanne W., Kevin J. Beiler, Marcus A, Bingham, Julie R. Deslippe, Leanne J. Philip, and François P. Teste. 'Mycorrhizal Networks: Mechanisms, Ecology and Modelling', *Fungal Biology Reviews*, Vol. 26 (2012): 39–60.

Simpson, George Gaylord. *Principles of Animal Taxonomy*, Columbia University Press and Oxford University Press, New York/ London, 1961.

Smellie Jnr., William. *The Philosophy of Natural History*, Charles Elliot, Edinburgh, 1790.

Spallanzani, Abbé Lazarro. *Dissertations Relative to the Natural History of Animals and Vegetables*, J. Murray, London, 1789.

Spilsbury, Richard and Louise. *The Life of Plants: Plant Reproduction*, Heinemann, Portsmouth NH, 2008.

Sprague, T. A. 'Plant Morphology in Albertus Magnus', *Bulletin of Miscellaneous Information (Royal Botanic Gardens, Kew)*, Vol. 1933 No. 9 (1933): 431–40.

Stafleu, Frans A. *Linnaeus and the Linnaeans: The Spreading of Their Ideas in Systematic Botany*, 1735–89, A Oosthoeks's Uitgeversmaatschappij, Utrecht, 1971.

Stannard, Jerry. 'Albertus Magnus and Medieval Herbalism', in J. Stannard, K. E. Stannard, and R. Kay, eds., *Pristina Medicamenta*, Ashgate, Aldershot, 1999.

Stepan, Nancy Leys. 'Race and Gender: The Role of Analogy in Science', *Isis*, Vol. 77, No. 2 (June 1986): 261–77.

Stephenson, Andrew G. and Robert I. Bertin. 'Male Competition, Female Choice, and Sexual Selection in Plants', in Leslie Real, ed., *Pollination Biology*, Academic Press, Orlando etc, 1983, 109–49.

Strasburger, Eduard. 'The Periodic Reduction of the Number of Chromosomes in the Life-History of Living Organisms', *Annals of Botany*, Vol. VIII, No. XXXI (September 1894): 281–316.

Svedelius, Nils. 'Alternation of Generations in Relation to Reduction Division', *Botanical Gazette*, Vol. 83, No. 4 (June 1927): 362–84.

Taiz, Lincoln and Lee Taiz. *Flora Unveiled: The Discovery and Denial of Sex in Plants*, Oxford University Press, Oxford, 2017.

Taiz, Lincoln, Eduardo Zeigler, Ian Max Møller and Angus Murphy. *Plant Physiology and Development*, sixth edition, Sinauer Associates/Oxford University Press, New York and Oxford, 2018.

Takabayashi, Junji and Marcel Dicke. 'Plant-carnivore Mutualism through Herbivore-induced Carnivore Attractants', *Trends in Plant Science*, Vol. 1, No. 4 (1996): 109–13.

Tassin, Jacques. *À quoi pensent les plants?*, Odile Jacob, Paris, 2016.

Tassin, Jacques. *Penser comme un arbre*, Odile Jacob, Paris, 2018.

Telewski, Frank W. 'A Unified Hypothesis of Mechanoperception in Plants', *American Journal of Botany*, Vol. 93, No. 10 (2006): 1466–76.

Theophrastus. *Enquiry into Plants*, Books 1–5, trans. Arthur Hort, Loeb/Harvard University Press, Cambridge MA and London, 1916.

Theophrastus. *Causes of Plants*, (*De Causis Plantarum*), trans. Benedict Einarson and George K. K. Link, Heinemann: London and Harvard University Press: Cambridge MA, 1976.

Thompson, Ken. *Darwin's Most Wonderful Plants: Darwin's Botany Today*, Profile Books, London, 2018.

Thornton, Robert John. *A New Illustration of the Sexual System of Linnaeus*, Volume 1, Printed for the author by T. Bensley, London, 1799.

Toepfer, Georg. 'Teleology and Its Constitutive Role for Biology as the Science of Organized Systems in Nature', *Studies in History and Philosophy of Biological and Biomedical Sciences*, Vol. 43 (2012): 113–19.

Tompkins, Peter and Christopher Bird. *The Secret Life of Plants: A Fascinating Account of the Physical, Emotional, and Spiritual Relations between Plants and Man*, Harper, New York, 1973.

Tournefort, Joseph Pitton de. *Elemens de Botanique, ou Methode Pour Connoître les Plantes*, 1694.

Tree, Isabella. *Wilding: The Return of Nature to a British Farm*, Picador, London, 2018.

Trewavas, Anthony. 'Aspects of Plant Intelligence', *Annals of Botany*, Vol. 92 (2003): 1–20.

Trewavas, Anthony. 'Reply to Alpi et al: Plant Neurobiology – Metaphors Have Value', *Trends in Plant Science*, Vol. 12, No. 6 (June 2007): 231–3.

Trewavas, Anthony. 'What Is Plant Behaviour?', *Plant, Cell and Environment*, Vol. 32 (2009): 606–16.

Trewavas, Anthony. *Plant Behaviour and Intelligence*, Oxford University Press, Oxford, 2014.

Trewavas, Anthony. 'The Foundations of Plant Intelligence', *Interface Focus*, Vol. 7, No. 3 (2017): 20160098.

Tudge, Colin. *The Secret Life of Trees*, Allen Lane, London, 2005.

Tuhiwai Smith, Linda. *Decolonizing Methodologies: Research and Indigenous Peoples*, Zed Books, London, 2012.

Turbayne, Colin Murray. *The Myth of Metaphor*, University of South Carolina Press, Columbia SC, 1970.

Turner, Dale. *This Is Not a Peace Pipe: Towards A Critical Indigenous Philosophy*, University of Toronto Press, Toronto, 2006.

Twetten, David and Steven Balder. 'Introduction', in Irven M. Resnick, ed., *A Companion to Albert the Great: Theology, Philosophy and the Sciences*, Brill, Leiden, 2013.

Ueda, Minoru and Yoko Nakamura. 'Metabolites Involved in Plant Movement and "Memory": Nyctinasty of Legumes and Trap Movement in the Venus Fly Trap', *Natural Product Reports*, Vol. 23 (2006): 548–57.

Vaillant, Sébastien. *Discours sur la structure des fleurs, leurs différences et l'usage de leurs parties*, Pierre Vander, Leiden, 1718.

Veits, Marine et al. 'Flowers Respond to Pollinator Sound within Minutes by Increasing Nectar Sugar Concentration', *Ecology Letters*, Vol. 22 (2019): 1483–92.

Vickers, Brian. 'Analogy Versus Identity: The Rejection of the Occult Symbolism 1580–1680', in Vickers, ed., *Occult and Scientific Mentalities in the Renaissance*, Cambridge University Press, Cambridge, 1984.

Viveiros de Castro, Eduardo. 'Cosmological Deixis and Amerindian Perspectivism', *The Journal of the Royal Anthropological Institute*, Vol. 4, No. 3 (September 1998): 469–88.

Viveiros de Castro, Eduardo. *Cannibal Metaphysics: For a Post-Structural Anthropology*, trans. Peter Skafish, University of Minnesota Press, Minneapolis and London, 2017 [2009].

Wakefield, Priscilla. *An Introduction to Botany in a Series of Familiar Letters*, Thomas Burnfide, Dublin, 1796.

Wall Kimmerer, Robin. *Gathering Moss: A Natural and Cultural History of Mosses*, Penguin, London, 2021 [2003].

Wall Kimmerer, Robin. *Braiding Sweetgrass: Indigenous Wisdom, Scientific Knowledge and the Teachings of Plants*, Penguin, London, 2013.

Webb, Colin J. 'David Graham Lloyd', *New Zealand Journal of Botany*, Vol. 45 (2007): 443–9.

Withering, William. *Botanical Arrangement of British Plants*, Volume 1, M. Swinney, Birmingham, 1788 [1776].

Wohlleben, Peter. *The Hidden Life of Trees: What They Feel, How They Communicate: Discoveries from a Secret World*, trans. Jane Billinghurst, William Collins, London, 2017 [2015].

Xu, Jianping. 'The Prevalence and Evolution of Sex in Microorganisms', *Genome*, Vol. 47, No. 5 (2004): 775–80.

Zalužanský, Adam. 'De sexu plantarum' [1592] in Holger Funk, 'Adam Zalužanský's "De sexu plantarum" (1592): An Early Pioneering Chapter on Plant Sexuality', *Archives of Natural History*, Vol. 40, No. 2 (2013): 244–56.

Zirkle, Conway. 'Introduction' to N Grew, *The Anatomy of Plants*, Johnson Reprint Corp., New York/London, 1965.

Zweifel, R. and F. Zeugin. 'Ultrasonic Acoustic Emissions in Drought-stressed Trees – More Than Signals from Cavitation?', *New Phytologist*, Vol. 179 (2008): 1070–9.

Filmography

Intelligent Trees. Dir. Julia Dordel and Guido Tölke, 2016.
Mother Trees Connect the Forest. Dir. Dan McKinney and Julia Dordel, 2011.

INDEX

Milton Keynes UK
Ingram Content Group UK Ltd.
UKHW022315191123
432893UK00002B/22